SEMICONDUCTOR DESIGN
AND IMPLEMENTATION ISSUES IN
INTEGRATED VEHICLE ELECTRONICS

Semiconductor Design and Implementation Issues in Integrated Vehicle Electronics

John H. Hall

Academica Press, LLC
Bethesda

Library of Congress Cataloging-in-Publication Data

Hall, John H. (John Haslet), 1932-
Semiconductor design and implementation issues in integrated vehicle electronics / John H. Hall.
    p. cm.
Includes bibliographical references and index.
ISBN 1-933146-01-X

1. Automotive computers. 2. Automobiles—Electronic equipment. 3. Integrated circuits—Reliability. I. Title.

    TL272.53.H35 2005
    629.2'7—dc22
                        2004022580

Editorial Inquiries:
Academica Press, LLC
7831 Woodmont Avenue, #381
Bethesda, MD 20814
Website: www.academicapress.com
To order: (650) 329-0685 phone and fax

Layout & printing by
Pensoft, Bulgaria

This book is dedicated to

John Michael, Jennifer, Jasmine, and MaryHelen

for their kindness and support.

Table of Contents

Illustrations

Preface

I would like to dedicate this book to my late friend and mentor, Dr. Jean Hoerni. As one of the "Fairchild Eight" semiconductor pioneers, Jean developed the planar technology that formed the basis of all integrated circuits. It is not well known, but Hoerni's invention of the planar process was a huge advancement in the reliability of semiconductors. This breakthrough allowed developers to build the complex devices discussed here.

Jean and I met because of an early problem related to the control of guided missiles. This problem was very similar to those experienced several years later in the design of semiconductor products for automobile applications.

In 1960 the Minuteman intercontinental ballistic missiles (ICBMs)—developed as the U.S.'s front-line nuclear capability—suffered a series of catastrophic malfunctions during tests as they exited the atmosphere. The cause of these failures became the focus of my work. I was able to isolate the cause of the failures to electrical interference that caused a computer malfunction. Later I was able to develop a system to eliminate these kinds of malfunctions, and it was through this development that I initially became associated with Hoerni.

This early work in semiconductor and computer design led to my interest in the ongoing problem of semiconductor designs for automobile applications. The problems encountered in building semiconductors for consumer vehicles are similar to those encountered in developing electronics for missiles and other aircraft. Techniques developed to ensure semiconductors will not malfunction during their intended use are often transferable from one type of vehicle to another. Hoerni's innovation and the experience I gained during the 10 years we worked together has been the main inspiration for my work during the last 40 years.

Hoerni's basic contribution to the automobile vehicle integrated circuit application allowed complex linear and digital applications to be constructed in a dense form. This permitted complex functions to be constructed in a small package with a very high degree of reliable operation, even in the severe environment experienced in vehicles. The environments experienced in automobile applications are extreme in that the temperatures go from many degrees below freezing to higher than boiling water. The parts are exposed to freezing rain as well as salty water environments and then the searing desert sun, sometimes over a very short time cycle.

In addition, they are exposed to intense surges of electromagnetic radiation generated both internally and externally. Hoerni's planar process has allowed manufacturers to build products that withstand all these extreme environmental factors with superb reliability. There are few contributions to science that exist in the world today that have had such a broad impact on the way that we experience and function in our lives.

But even fundamentally good scientific advancement is not always implemented using sound design techniques. Therefore, the problem of semiconductor systems failing during intended use has been an ongoing focus of my career time and time again. Quite often systems are developed using the best skills of qualified designers, yet they fail during their intended use. The causes of these failures are often mysterious and difficult to detect and repair.

In 2003, a friend of mine was driving a 1998 Ford Explorer during a snowstorm on the East Coast. While proceeding up a snow-covered major interstate highway at a steady 40 mph (64 kilometers/hour), the vehicle's front wheels began to spin faster than the back ones, an apparent malfunction of its four-wheel-drive torque distribution system. The front of the SUV lost traction, the vehicle slid sideways and went under a tractor-trailer. My friend miraculously was not hurt, but the SUV was totaled. The question of what would cause this malfunction occupied my mind for some time and led me to do some research on the possible causes of this mishap.

Four-wheel drive presents many complex engineering challenges. During the Second World War, Willys Jeeps deliberately geared the front wheels so they would spin faster than the back ones. This allowed the vehicles to climb hills more effectively, but also meant the four-wheel drive could not be engaged on hard, dry pavement. The Willys and its contemporaries were thus optimized for some environments, but not for others. The capabilities and limitations of these systems were obvious to users.

Fast-forward to the age of sensors, processors, and electronically controlled drive trains. In the search for the perfect four-wheel-drive vehicle, modern SUV designers have developed complex ways of varying the speed of front wheels so they can be engaged on a variety of surfaces effectively. Key to this design approach are sensors, microprocessors and software programs that manipulate the speed of each individual wheel in obvious and non-obvious ways in an effort to make four-wheel-drive vehicles easier to drive in difficult conditions.

New systems add obvious capabilities—the new, more advanced four-wheel-drive in vehicles can be used by drivers with no special experience, greatly expanding the market for them. These capabilities are accompanied by limitations that are less obvious to users and even to the automakers themselves.

I have found over the past 40 years that the implementation of new vehicle electronic features is an odd dichotomy. On one hand, obvious electronics such as electronic car alarms were available on an after-market basis a decade before automakers began including them in their vehicles. But on the other hand, automakers often add other features that appear to be less than necessary, and driven by marketing, rather than a desire for safer and more effective vehicle operation.

Furthermore, connecting dozens of sensors and disparate systems such as radios, windshield wipers, transfer cases and power windows into a single processor (computer)—using multiple processors would cost much more—is a risky engineering approach that will often, if not always, lead to significant failures in use. Yet this has been done for millions of vehicles. Owners have complained of problems such as wipers coming on without the switch being operated, but failed to notice less obvious operational malfunctions, such as transfer case failures.

Testing will often produce a false sense of design security. In the case of the Minuteman Missile, failures happened while flying through clouds in California, and since it was often tested in Florida on clear days the system generally seemed to work. It was not easy to understand that the weather could cause a computer to malfunction. Had testing been done methodically in all weather conditions, the problem would have been obvious.

The story of automotive vehicle failures is just as complex and often relates to integration of various electronic systems. Failures such as windshield wipers coming on for no apparent reason are difficult to connect with the malfunction of the power train by studying schematics. However, once the nature of the failure is recognized, it is relatively easy to connect the two malfunctions.

Introduction

Many previous articles and some larger works by competent authors were dedicated to the important subject of complex vehicle electronics. This effort fills what I believe to be a significant void. Other authors, drawing on a wealth of information provided by automakers and electronics suppliers, have illustrated the payoffs in advanced features. While these works have added a great deal to understanding of this subject, a more complete comprehension of design issues is needed.

The key concept the reader must grasp is that a microprocessor built to control vehicle actions can fail in ways that are difficult to predict. Logic sequences designers believe are failsafe—such as ones preventing shifting into four-wheel drive low at highway speeds—are by no means as secure as automakers pretend they are. Sensors sometimes send erroneous signals, tricking the processor into ordering a false action. The processor itself may fail temporarily in a variety of ways that cause it to order an unrequested action.

Often this is just annoying to drivers. Pontiac Grand Prix owners recently found that when they started their cars, the correct date appeared on the computer display, but the day of the week was incorrect. GM admits it is a programming error related somehow to leap year, but at the time I concluded my work on this book, the automaker had not announced a solution.

Readers should keep this example in mind as I describe the many ways marginal designs cause vehicle electronics to fail. Having developed the first electronic wristwatch to have the day/date function some 30 years ago, I am not climbing too far out on a limb when I state having a date and day mismatch is a rather obvious mistake.

The development of new, complex vehicle functions is much more difficult, and we can conclude objectively that these functions are more likely to contain errors and to fail in unexpected ways.

When vehicle electronics fail in obvious ways, it makes for good newspaper fodder. This recent USA Today article recounts one set of problems in excruciating detail:[1]

---

[1] O'Donnell, Jayne, "Tech Advances Make for High-Priced, High-Class Headaches," *USA Today*, 11 November 2003.

A California salesman says he had owned his 2003 BMW 7-series a few weeks in the spring when it stopped starting. When he hit highway speeds one evening after it was serviced, compact discs spit out at his passenger and the engine began sputtering and lurching until it died at the side of the road.

The problem: The computers running the state-of-the-art electronics in his $80,000 car were full of bugs. The car owner, who can't be identified because BMW insisted he sign a confidentiality statement when it bought back his car, isn't alone. Thanks to the latest electronics, cars can tell you the pressure in each tire, display stock quotes or give directions to the nearest Italian restaurant. But the complex computer systems required to do all that have broken down on hundreds, perhaps thousands, of luxury vehicles, wreaking havoc on the lives of their owners.

The defects such as those described above are obvious to any driver. More dangerous are problems that affect vehicle roadworthiness without providing warning signals or performance indications, or those that happen suddenly.

Vehicle electronics system failures are caused sometimes by the process forced on the engineers who develop the devices. These systems are usually incremental modifications to legacy systems rather than being designed from scratch, so the design is a compromise of the old and the new. Furthermore, engineers are rewarded for reducing production costs, so the cheapest possible materials and design deemed capable of performing the required tasks are used.

I ask the reader to be patient with some of the more technical descriptions. Determining the likelihood of semiconductor failure requires a thorough understanding of the physics of semiconductors and their operation in the automotive environment. The flaws, in these cases, are often measured in microns (millionth of a meter). My purpose is not to criticize, or to argue that developing innovative vehicle electronics should be discouraged. Rather, I describe examples of design flaws and reasonable techniques to reduce the probability of system failure.

The technology used in modern integrated circuits—used not only in the modules described here but also in most electronic systems—is based upon developments I made decades ago at Union Carbide's semiconductor operation, Intersil, and Micro Power Systems.

I also have over 10 years of experience related to the issues raised by vehicle systems, especially in devising techniques to make electronics more resistant to environmental factors. This includes the development of the first radiation-hardened computer for a classified Defense Department program. I mention this past work so readers might have some confidence in the solutions I suggest.

The book combines discussions of semiconductors, electronics systems, and vehicles in a way that is hopefully useful for persons developing and evaluating these technologies. Piles of recall notices, technical service bulletins, and user complaints serve as testaments that too often designers fall far short in providing consumers with dependable vehicle electronic systems.

The primary example studied for this book was the Ford Explorer. Prior to beginning research for this effort, I had little interest in, or knowledge of, SUVs, my preference being along the lines of Porsches and Buick Grand Nationals. But in deciding upon a system to use as a primary case study, I reviewed marketing literature from a wide variety of vehicle manufacturers. I found that to a significant extent, Ford placed advanced electronic-based features at the center of its marketing efforts. This was particularly evident for the Explorer line of sports utility vehicles. These highly imaginative and unusual systems promised to be useful in illustrating the issues that arise in developing vehicle electronic systems, and I was not disappointed.

1. Vehicle Electronics Background

Automobile manufacturing presents one of the most difficult design challenges faced by any type of supplier. Every part of the car—from carpets to windshields to door knobs—receive higher stress than in most any other consumer market area and must be designed to withstand thousands of hours of use. And since automakers are among the most aggressive companies in squeezing costs out of components, suppliers are placed in difficult and occasionally untenable positions.

Some background, from the points of view of electronics suppliers to auto makers, is useful at this point, since their business environment frames the design decisions they make. The elating prospect of selling large numbers of chips to a carmaker is tempered by the reality that these are among the toughest customers in the semiconductor business. Profit margins in this area are razor thin, design specifications highly demanding, and expected production runs often are reduced or ended without warning.

Some older readers might recall that in the earlier generation of cars AM radios had a big problem with static caused by the car operation. Static electricity discharge generated by the rolling of the tires and the wheels was a big problem. The ignition noise caused by the spark generating system was another large source of noise. The brush noise caused by the DC generator also was a great source of interference, and many cars had a strap that dragged behind the car onto the ground to release internally generated electrical charges. There were a number of filter designs incorporated into the radio that were unusual due to the wide bandwidth of the noise. One of these was a large flat plate, which was designed to produce a capacitor that had an extremely low inductance because it was built into the metal chassis of the radio. This was used in conjunction with some inductors to filter the power supply so that it would not produce noise into the radio.

Producing a noise-free automobile installation was a challenging task in the 1940s and 1950s. As a result there has been a continuing effort in the construction of automobiles to reduce the noise that they generate in order to make the radios work properly. This work has matured very well.

In 1964, I helped to start a company named Intersil with Jean Hoerni. The original idea for forming the company was to develop low power consumption integrated circuits for timekeeping applications for the Swiss watch and clock industry. The Swiss provided the financing and Hoerni and I provided the technology. One of our timekeeping applications was an automobile clock.[2]

In the 1960s, most automobile clocks used a mechanical clock movement with an electric winding mechanism. When the clock ran down, the extended e-ring closed an electrical contact, activating an electrical solenoid that mechanically re-wound the spring and kept the clock running.

The problem with this clock was that the electrical contacts used had a finite lifetime, and as a result the clocks always failed before the useful lifetime of the automobile was over. The early failure point in a car was its clock. For this reason, there was a great interest in an electronic automobile clock that had a long lifetime. Addressing this problem gave me an early experience in the application of electronics to automobiles at a time when the only other electronic device in the car was the radio. It was at this time that I learned about the severe conditions that one would encounter when developing for automotive applications.

One requirement faced by auto electronics suppliers was that the product must be very low cost to produce. Another was low power consumption, so that it would not run the battery down during long storage times since it ran all of the time. Still another was increased tolerance of temperature extremes and mechanical shock. One of the most difficult requirements, however, was the ability to withstand severe electrical transient voltage swings that appeared in the electrical system. These were of very high voltage (thousands of volts) and high energy content, capable of destroying any normal semiconductor electronic device connected to the electrical system.

As a result of this problem it was necessary to add high-power protection devices external to the clock integrated circuit (IC) to protect it from these high-energy transients. The eventual solution to the problem was a quartz crystal frequency reference, an integrated circuit frequency divider, which operated a low power consumption motor that would step at one step per second, and drive a set of clock hands as a display.

Other projects also needed electronics for harsh environments. I encountered this need again while working for the aeronautical computer design group at Honeywell Corporation in Florida. They had a contract to develop the aircraft computer for the SR-71 high altitude spy plane. At that time most compact military computer designs had failed, and as a result Honeywell was very concerned that such an important project be successful. This computer was to be an all-IC computer, and, as a result, the logic signal levels would be smaller than those previously used for

---

[2] Ironically, our Swiss backers refused to build the electronic watch after we had produced the necessary technology, saying they would rather keep their paid-for mechanical watch factories running than build new ones for semiconductors. Seiko bought the technology instead and used this to transform itself from a small manufacturer of mechanical watches into a world-wide electronics manufacturer.

factory computers. This fact would make the computer more susceptible to electrical noise than discrete logic signals designs, which used higher logic voltages.

We used every noise reduction method known at that time and our work went very well, and the result was a very reliable and stable operating computer, which was successfully produced for over 10 years. A similar system designed by another group did fail as a result of noise problems, but after I helped in the redesign, it was successful.

Note that most of the early computer and system component designs utilized some form of bipolar transistor design. This technology has very high speed and has good noise rejection, but has moderate-to-high power consumption. As systems and computers became more complex and densely packed, the technology was changed to P Channel MOS (metal oxide semiconductor) technology. This technology was used in early calculators and then the Intel type microprocessor chips, but its speed was not as fast as bipolar technology. The power consumption and internal heat generated did not allow bipolar technology to be used for complex one-chip systems. Later, CMOS (complementary-symmetry metal oxide semiconductor) technology was used for calculator and microprocessor manufacturing because it had lower power consumption and a high speed of operation.

In 1968, I pioneered the use of CMOS technology for calculator, watch, and memory applications. Due to my extensive work with CMOS technology, I became very aware of its susceptibility to transient electrical signals on its inputs, outputs, and power supply lines. This pointed to the need for protection of these terminals, as well as careful system design to avoid malfunctions. Over a decade later, this technology was adopted by Motorola, Intel, and Texas Instruments, as well as many Japanese companies. Many users experienced problems with the susceptibility of the technology to electrical transients because designers failed to implement proper protection.

My work in CMOS found immediate application in vehicle electronics. The next project resulted from a government requirement in the early 1970s for a seat belt alarm that would sound an alarm if a driver drove off without connecting the belt. The car companies asked their suppliers to issue development orders to several integrated circuit manufacturing companies to develop a set of ICs that would sense if the seat belts were connected, if someone was sitting in the seat and the car was started, and if the belt was not fastened. It came as a shock to the integrated circuit designers how severe the electrical transients were on the vehicle power lines, and a big problem was how to protect against them.

A protection system was provided outside of the IC, and an internal protection power diode was included that was so large it took up 25 percent of the IC area. At this time, my company, Micro Power Systems, received an order from the Stewart Warner Company, a large automotive supplier, to develop these devices.

Government regulation changes complicate the already volatile automobile market. The U.S. Government mandated the seat belt interlocks, and it seemed like an ideal opportunity for my company. But the Government then issued a waiver for the seat belt circuit, and Ford, one of the Stewart Warner Company's main clients, not wanting to be the only company producing vehicles with a less-than-popular feature, reasonably ended the project.

Designing a chip from scratch for a specific vehicle model, even one with a reasonably large production run, is a risky project for any semiconductor manufacturer, particularly given the enormous downward price pressure carmakers put on all their suppliers. Chipmakers must get the design correct early in the process, as costly redesign and new manufacturing tooling will wipe out any profits the project might have had. Adapting existing chips is often key to making these projects viable.

A few years later the government required automakers to reduce vehicle gasoline consumption and decided to rely on advanced electronics. In the case of the Ford Company, they assigned the job to one of their larger suppliers, The Essex Wire and Cable Company. Essex subcontracted to Analog Devices, and Micro Power Systems (my company). Analog Devices had to develop six ICs within six months to measure the exhaust gas and other signals, compute the necessary carburetor setting, and drive a motor that adjusted the carburetor metering using an electric motor.

By this time, high-voltage electrical transients had been reduced somewhat because automakers changed from Direct Current generators and voltage regulators to an Alternating Current generator with a built-in regulation system that produced fewer transients. Improvements were also made in the starter motor system, as well as the spark plug ignition system.

The unfortunate part of the project was that the Government relented on its demand for the strict requirements in the model year, and, as a result, Ford dropped all of their requirements. Essex as well dropped their work and its relationship with Analog Devices. The exercise, however, gave us a familiarity with the environmental requirements, which involved working with high voltage transients and noisy conditions.

The combination of CMOS development and more friendly auto power generation systems made more advanced vehicle electronics possible, but designers still faced a minefield of potential problems.

The most serious problem with CMOS technology is that after it has been exposed to a transient higher than its power supply voltages on either an input or output (in current state-of-the-art technology two volts or less), the whole chip can turn on (become a short circuit) in the form of a silicon-controlled rectifier. This state is referred to as "latch-up." This activity can either damage the chip permanently or cause it to operate in an erratic manner for a period of time with no permanent damage.

Please note that in most designs, when the computer chips are operating properly they put out a computer "word" (group of signals) onto a multi-wired bus. This word will then be interpreted by one of a number of receiving units at the other end of the bus to perform some discrete function. The computer output word could cause the windshield wipers to come on, the windows to roll down or the transmission to shift. When the computer chip is latched up by an over voltage transient on either the power supply lines or any of the input or output lines, an inappropriate output word can appear on the control bus, which can result in inappropriate activity.

Problems can come from unexpected sources. For example, a source of interfering electrical signals can come from cell phone radio frequency signals, which demonstrated itself during the early development of automatic braking systems. Brake sensors were located at each of the four wheels, and each sensor had to transmit signals to a central location in the engine compartment. The interconnecting wires formed an antenna, which would allow cell phone signals to interfere with the braking systems. Proper engineering techniques such as shielding solved this problem.

This case illustrates that the major problem for complex vehicle electronics is that the number of conditions that can affect an operating system is so large it is virtually impossible to build a system that is impervious to all potential interfering conditions. For this reason, many systems have come onto the market affected by electrical glitches, and you will see cars and trucks that have had a large number of engineering changes incorporated into them after being sent to market. Sometimes these changes have manifested themselves through massive recalls, but more often it appears that they are incorporated quietly by the dealers when the vehicle is brought in for some complaint.

The electrical interference signals are only one of the problems that have been experienced by the automakers. Computer software has been a massive problem because it frequently causes improper operation of various functions within the automotive system. If you study the history of some cars you will find that there are a huge number of software upgrades that are installed in the computer systems either during a recall or during routine maintenance by the dealer.

This problem is further complicated by poorly designed sensors that disrupt the system operation. A good example of this was the speed sensors for the automatic transmission system of one brand of cars. The failure of this sensor would cause an erratic operation in the transmission and the four-wheel drive system, which frequently destroyed either the transmission or the four-wheel drive transfer case. Unfortunately, in this case the dealer and the manufacturer usually denied any responsibility and required the customer to replace the damaged components. As the electronic systems become responsible for more and more functions—such as controlling the gas pedal speed control and electronic steering control—failures like those experienced with the speed sensors could result in catastrophic consequences.

Still another problem area is that of the interconnection of the wiring system, particularly electrical connections to sensors and electronic control devices. The connections are exposed to large amounts of road hazards such as rain, debris, snow, ice, salt, and large temperature variations. A number of failures in systems have occurred as a result of connectors that were required to attach the system electronics to transmissions, speed sensors, and transfer cases. In some instances, salty water that permeated the rubber seals on these devices caused the electrical connections to corrode, resulting in sometimes severe failure conditions of major components. The failure of major components could cause difficult-to-diagnose accidents.

In addition to being susceptible to failure, auto design features are subject to the fickle whims of consumers. For example, in 1978, I was producing chips for Cadillac that I had designed for an integrated CB radio system. This was a rather advanced project, incorporating many features in a single chip. But as some might remember, the market for CB radios collapsed suddenly, ending both my business with Cadillac as well as my aftermarket CB sales. Overnight I went from building many of these chips to virtually none.

2. Semiconductor Engineering Background

The Physics of Semiconductor Failure Modes

The ideal environment for electronic systems is a cool, humidity-controlled room free from static electricity or any sort of electrical interference. As described in the previous chapter, the exact opposite of this perfect environment is an automobile. The weather places electronics under significant stress in obvious ways. Transient voltages generated by a variety of sources and transmitted in different ways are a constant threat to vehicle electronics. These threats appear often in non-obvious ways.

The intent of this section is to explain how vehicle electronic systems are affected by energy and how they can be protected to a large extent against damage. Since microcircuits are the most vulnerable parts of modern electronic systems, much of my work has focused on the failure mechanisms of these devices and how to protect them against the high internal voltages.

This work also focuses on the way that electronic systems are built today, and will be in the future—with integrated circuits utilizing complementary-symmetry metal oxide semiconductor technology (CMOS). This technology has limited reliability in an unstable electrical environment since it contains a fundamental failure mechanism called "latch-up" that make the majority of Commercial Off The Shelf (COTS) equipment very susceptible to the energies generated by electrical transients and electromagnetic environments. Since IC technology is evolving to ever-smaller feature sizes (construction line widths), the susceptibility of ICs to latch-up failure increases with this miniaturization. Also, the susceptibility to failure from internal systems voltages with COTS equipment will increase as time passes.

This section is important for the reader even if he or she has trouble following the intricacies of semiconductor design and operation. This description of semiconductor failure will give the reader a deeper appreciation for what can go wrong in vehicle electronics systems.

Catastrophic Failure Mechanisms in Semiconductor ICs

There are six major failure mechanisms in Integrated Circuits:
1. Localized high temperatures;
2. Localized dielectric breakdown;
3. Metal corrosion;
4. Electro-migration of the metalization;
5. Surface inversion caused by ionized contamination in the passivating silicon oxide layers, and;
6. CMOS latch-up.

Only three of these are affected by externally applied electrical fields: localized high temperatures, localized dielectric breakdown, and CMOS latch-up. The high-induced currents caused by sources in vehicles or outside of them could create all three of these problems.

Failure by localized heating of the silicon surface results from an excessive current flowing through a (P-N) junction formed in a particular portion of an integrated circuit. Such junctions usually form transistors or protection diodes. The areas of these P-N junctions are usually small. If an excessive current flows in one of them, either in a forward or reverse direction, internal resistance can generate a high temperature. If the current flow is in the reverse direction, a higher temperature can be generated within the junction area as a result of the reverse current heating effect. The local temperature can easily reach 700^0 to $1,000^0$ centigrade.

Most electrical P-N junctions of interest—such as an input or output transistor in either a bipolar transistor or a MOS transistor integrated circuit—have a local contact to a metal connection formed from aluminum or an aluminum alloy. This aluminum connection deteriorates above 500^0 C, diffusing rapidly into the silicon and then becoming molten when above 600^0 C. This molten aluminum moves rapidly through the P-N junction, causing a short circuit. Once the junction becomes short circuited, the function associated with that junction fails. In addition to this failure, the metal connection of the junction is often consumed by the molten aluminum and is destroyed.

This type of failure is frequently seen at the input protection diode of an MOS IC or on an output driver transistor. An overvoltage and current at either terminal will cause this type of failure. This failure also occurs frequently on an input transistor of a bipolar logic IC as the result of an overvoltage on one of the input terminals. Bipolar transistors have a similar failure mechanism because of their small area emitters.

Localized heating also causes failures with IC inputs, outputs, and power supply connections. The surface of these contact pads are connected to the pins of the IC package with gold or aluminum wires one-thousandth of an inch in diameter. An electrical current of over one amp will cause resistance heating of the wire, with the resulting localized heating causing the wires to fuse. Either internal or external factors can cause such excessive current at this location in the IC.

With respect to localized dielectric breakdown, failure would happen in two areas: 1) transistor gate regions; and 2) the interlayer dielectric material between the conductive (metal) interconnect layers. Gate dielectric failure in MOS transistors is caused when the dielectric, usually silicon dioxide, is stressed above the dielectric electrical strength in its weakest area. Dielectric strength is measured in volts per millimeter, and as gate dielectric thickness decreases, its ability to sustain a voltage between its surfaces decreases. Often the oxide film is not perfectly uniform in thickness and so it can only sustain the voltage capability of its weakest point.

A failure in the dielectric results in a high-density current flow in a localized area, which generates an intense heat, leading to the failure of the dielectric in a small, localized area. Heat generated at these points forces the temperature in the defect to rise to over $1,000^0$ C. This causes an aperture to form in the dielectric and the silicon or metal electrodes to melt, flowing into the failure site and short circuiting the electrodes.

This electrode short circuit can cause further destruction in the IC. When the electrodes are the gate element of a MOS transistor, the gate electrode becomes shorted to one of the power supply connections and the transistor fails. When the electrodes are two layers of metal interconnect and the dielectric between them fails, two inappropriate metal lines become shorted together and a circuit malfunction becomes permanent. This malfunction can be catastrophic, partially debilitating, or invisible. In many cases it causes latent defects.

Large-scale MOS integrated circuits are tested, handled, and packaged very carefully because large electrostatic voltages on equipment can cause failures due to these potentially fatal failure mechanisms. Fabrication and design procedures seek to prevent inadvertent high voltages resulting from the build up of high electrostatic charge on humans and machinery caused by friction on dielectric materials. Electrostatic charge is characterized by a high voltage with a very limited capacity to supply current.

The IC inputs are protected by circuits incorporated into the IC itself after their final installation in equipment. These protection circuits consist primarily of diode clamps that drain the electrical potential on an input to the power supplies in the event that the input voltage exceeds the power supply voltages by more than 0.7 volts. These devices utilize a current-limiting resistor and rely on the assumption that the energy in such an overvoltage is considered to be very low. Such is the case in an electrostatic charge on a human body.

The output pins of the IC are connected usually to the drains of large area N Channel and P Channel transistors, a design technique that forms large area protection diodes to the power supplies. Usually, no further protection is provided.

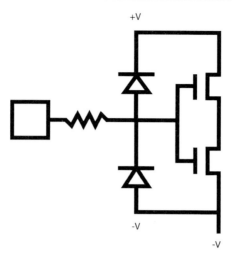

Fig. 1. **CMOS Input Circuit Showing Protection Diodes**

In bipolar ICs, the input currents are limited by the low energy of the electrostatic voltage and internal resistance, so no protection is required. The output transistors are of a large area so that no protection is needed due to the low energy of the usual electrostatic charge.

A practical problem is that as product developers strive to bring to market more complex and faster products, the junction area of the protection diodes becomes smaller and the gate area of the IC becomes thinner and more fragile. The total interconnect area becomes larger, has more layers, and becomes more fragile, giving rise to more interconnect shorts.

The trend with time due to smaller feature sizes is toward a lower potential for survivability of systems exposed to excess voltage. In addition to these trends, the operating power supply voltages are moving toward three volts from five volts and to 1.5 volts, which will make future products even more vulnerable to excess voltage.

Other Failure Mechanisms in CMOS Large Scale Integrated Circuits

CMOS latch-up is by far the most prominent failure mechanism in system equipment utilizing CMOS LSI (large-scale integration). A malfunction can be expected if any input or output pin voltage is raised above the power supply voltages, and, as a result, causing current flow into the chip. When this occurs, a latch-up condition can be established within the IC, causing the power supply connections or chip interconnects to fuse.

On the other hand, because the latch-up currents are often localized to a limited area of the chip and the failure current is limited by internal resistance, the failed portion of the circuit can often completely recover once the power-supply holding current is removed. This is accomplished by disconnecting the power supply voltage for a short time. In another type of failure mode, the function of a portion of the chip is disturbed as a result of the injection of high-energy carriers into high sensitivity chip areas.

This type of failure can be returned to normal operation through an unpredictable time and temperature controlled relaxation process. It involves the injection of hot carriers into the silicon dioxide surface passivation layer. These hot carriers become trapped in the passivating layer, causing a charge build-up to form. The charge in the oxide causes an inversion layer to form in the underlying silicon, which can cause high leakage currents to flow or cause short circuit paths to form between adjacent transistors. These failures are more prevalent in modern CMOS technology using LOCOS (local oxidation of silicon) oxide types of transistor isolation.

A third type of failure mechanism in CMOS is the gate rupture mechanism. Gate oxide thickness becomes increasingly thinner as manufacturer's scale their technology to ever smaller dimensions to achieve higher speeds and circuit densities. Today the gate oxides are about 25 angstroms thick, and decreasing toward 10 angstroms. In the past, internal transistors have been protected by circuit internal protection diodes, which limit the input voltage to six volts for five volt ICs.

Currently, external diodes provide additional protection and are in the form of external Zener protection diodes mounted on the application circuit board to improve the reliability of those IC applications exposed to high electrostatic discharges. Such external protection "clamps" an overvoltage appearing on a system input or output to ground so that it stays within five volts.

Protecting state-of-the-art and next-generation ICs—at 3.3-volt and 1.5-volt operation—becomes more difficult because there are no two- or three-volt Zener diodes.

Both input and output terminals of CMOS ICs are connected to long printed circuit patterns on the application interconnect printed circuit boards. These input and output pins of the central processor, for example, communicate to and from memory and various input and output devices via highly inductive printed circuit traces. These traces develop high-energy, high-voltage signals when exposed to high (RF) radio frequency fields or other sources of overvoltages.

You might have noticed instructions while filling your gas tank telling you not to use your cell phone while pumping gas. Radio frequency-producing devices such as cell phones—when used at very close ranges—can have a severe effect on electronics. There are many sources of this type of energy, and it can do a great deal of damage to vehicle electronics.

Radio frequency energy entering the terminal housing via these ports would then interact with the interconnection traces on the computer mother board, causing high voltage spikes to appear directly on the computer and associated IC input/ output and control pins resulting in IC failures.

The point of this discussion is that CMOS will fail if implemented improperly. A great deal of excellent work has been done in the area of protecting CMOS-based circuits from suffering the failure mechanisms outlined above. In Chapter 6, I discuss some of my work, and that of others, in the area of CMOS protection.

3. Current Vehicle Electronics Design Techniques and Issues

Computer and electronic production is a diverse industry, ranging from today's disposable consumer electronics to the super high reliability and longevity of military hardware and space flight. The differences between the various quality levels have existed for decades, with steps being taken towards improvements and advancements in materials and technology by all in the field. It may be surprising, but at the higher quality level, the processes and materials are less flexible, and therefore the older technologies prevail. This is based somewhat on the stringency of the requirements, and on the philosophy that proven technology is safer and more reliable.

These quality and production issues come clearly into focus when considering vehicle electronics, as vehicle makers have been relentless in incorporating electronics into their products. According to The Alliance of Automobile Manufacturers:[3]

> Today cars, trucks, minivans and SUVs incorporate a high level of computer technology to meet the needs of consumers for safety, environmental, communications and even entertainment needs. In fact, electronics now control more than 86 percent of all systems in a typical vehicle.
>
> And while today's vehicles boast an impressive amount of computing power, the future promises a dazzling array of technology that may manage driver commutes, help avoid accidents and even conduct a body scan to put people in the best driving position. As powerful computers and computer chips become smaller, they play a larger role in changing the face of modern mobility.

The technological advancements that make today's ordinary consumer electronics cheap and plentiful, simply will not withstand harsh environments. As discussed before, the modern vehicle is a harsh environment by nature. The method for manufacturing highly reliability electronics, from a hardware stand-

---

[3] "Today's Automobile: A Computer on Wheel," *The Alliance of Automobile Manufacturers* news release, 22 March 2004.

point, is a twofold approach. First, the devices and assemblies are constructed out of specific material using proven processes. Second, the devices undergo extensive stress testing to weed out the weak features.

Manufacturers have learned what conditions wear out equipment. Some are obvious, like mechanical shock and vibration. Heat and cold extremes can contribute directly to failure, particularly at the component level. The change from one to another, or more accurately, temperature cycles, will degrade almost anything given enough exposure. Moisture, salt, dirt, and other foreign material can cause corrosion, and are particularly damaging to electronics because these materials are conductive as well as corrosive. Moisture causes damage for these reasons, as well as supporting organic growth, such as fungus. In the daily routine of vehicle operation, all of these are present.

The various methods to control the effects of these factors are numerous and fairly straightforward. Robust, rigid assemblies will withstand shock and vibration. Quality materials and an understanding of their thermal coefficient of expansion will control the degradation from temperature cycles. Issues of temperature exposure are controlled through design, and component selection. Corrosion and moisture damage is controlled through material selection, and sealing moisture and foreign material out. Even the most basic materials (dirt, salt, etc.) can become conductive when wet—even dead insects!

All vehicles are not created equal, and neither are the electronic modules that control them. The old adage that you get what you pay for in a vehicle does not necessarily hold true when it comes to electronics. You may get more features and functions, but this does not correlate to reliability. In the standard reliability calculations for electronic assemblies, simple circuits almost always score higher than complex ones. It is the design and construction of the assembly that is most important, and this applies all the way down to the design and construction of the components themselves. For example, gold contacts corrode less and conduct better than tin. Ceramic and metal packages have a higher tolerance for temperature cycles than plastic. Hermetic devices and sealed assemblies last longer than non-hermetic or exposed circuitry. Cast aluminum cases are less subject to vibration and mechanical shock than sheet metal or plastic cases.

There is one final point to consider when discussing module reliability from a design and construction standpoint. It is often difficult to know who actually designed or built the module. All the major U.S. manufacturers share technology with foreign manufacturers, and they all have factories all over the world. The community of electronic components and assemblies is truly a global one. The electronics industry is well immersed in the current quality philosophy of utilizing Quality Management Systems, or QMS. The well-known quality standard ISO 9000 is a QMS. It dictates the methodology of engineering and manufacturing control that, if followed, will produce a quality product. All of the electronic communities (automotive, medical, aerospace, military, consumer, etc.) require adherence to a QMS.

Each community embraces a specified set of requirements, or standards, applicable to their industry. These standards range from the basic QMS, to the individual requirements of the applicable performance specifications for components and assemblies that make up the final product. These requirements flow down to even the basic component manufacturers who make the transistors and ICs of the system modules. No system is perfect, however, and the performance standards all deal with the tolerances issues of multi-tiered manufacturing. By the time a module is installed in a vehicle, literally dozens of manufacturers will have been involved. The number of vehicles produced in a year, multiplied by the number of modules, assemblies, and components involved results in a huge number of assembly operations, interconnects, and processes affecting the final products... the vehicles we depend on every day. It becomes obvious then that even a 99 percent accuracy rate is insufficient in this scenario.

To better understand the wide range of technology and the subsequent design and construction of vehicle electronic modules, a variety of computer modules were disassembled and analyzed. This is a physical analysis primarily, as schematic diagrams are generally not available outside the manufacturer. One can, however, develop a basic understanding of the complexity of the circuitry through an understanding of electronic component types, and device and interconnect counts. The results of these analyses are as follows:

Subject Computer: ECU for 1989 Chevrolet S-10 Pickup, 2.5 Liter

Mechanical Aspects

This electronic control unit (ECU) is housed in a stamped and folded sheet metal case. The case offers no moisture protection, and has small openings at the corners that dirt and insects can get through. There is an access cover, with a gasket, to allow access to a device that is mounted in a removable carrier. The case is closed with hex-head fasteners with tapered ends. These fasteners could tap their own threads; however the condition of the metal around the holes indicates they were tapped. Tapping is preferable as sheet metal screws generate conductive aluminum particles.

Inside, the computer uses a single printed circuit board that is bolted to the lid using an internal aluminum structure, that also serves as the heat sinks for the output drivers. The drivers are mounted to the aluminum bars and are held in place with a plastic fixture.

Design

The circuit board is small, about 4.5" x 5". There are 8 ICs, some surface mount and some through-hole. The passive components are also a mixture of surface mount and through-hole technology. The PCB (printed circuit board) is double sided, with components on both sides of the PCB. The integrated circuits are all in plastic packages (commercial grade) and the size and pin counts indicate less sophisticated devices. All of the ICs have the marking intact. The soldering is uniform and well done.

Production Processes

The assembly indicates standard commercial processes. In this case, the PCB is fully assembled, including the heat sink/mounting hardware and then conformal coated using a spray/mask process. This is done to protect the PCB and components from moisture and foreign material. There is an edge connector on the back edge of the board, indicating on-board programming capability, but no corresponding opening in the case. This indicates the programming is done during assembly, and that there is no need to program after assembly is completed. The PCB assembly does not fill the case as with most computers. This would perhaps indicate the use of standard assemblies for the PCB and the case.

The removable carrier is 3.5" long, and holds a 26-pin EPROM (Electronic Programmable Read-Only Memory) and two other ICs. This is where the program, or software, resides. The ICs are mounted directly on the carrier, so no circuit board is used. The module mounts with a 65-pin connector with locking tabs to keep it secure.

The interface to the vehicle is a single ganged connector that accepts two plugs from the harness. The body of the connector is plastic and the pins are tin plated. The connector shell has locking tabs to keep everything together. This construction is consistent with a standard automotive grade connector.

Fig. 2. **ECU for 1989 S-10 Pickup**
The case for this Chevy S-10 module has openings that let in dirt and insects.

Comments

Compared to the other GM computers, and the Ford computers, this unit is much simpler. Although it appears cheaply made because of the case, it was built with good commercial practices.

Another interesting note is the tag on the case that states it is a remanufactured unit. The circuit board shows no sign of rework or repairs.

Fig. 3. **ECU for 1989 S-10 Pickup**
The printed circuit board for this module is doubled sided.

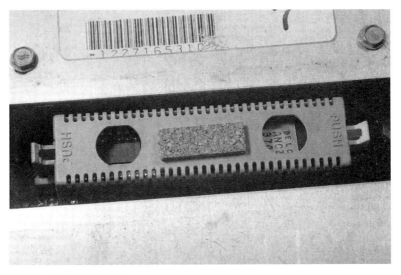

Fig. 4. **ECU for 1989 S-10 Pickup**
The removable carrier holds a 26-pin EPROM and two other integrated circuits.

Subject Computers: ECU for 1999 Chevrolet Z 7 Liter; 1999 Chevrolet 5.4 Liter V8 with Auto; 1997 Chevrolet S-10 Truck

Mechanical Aspects

These computers are housed in a three-part case. The bottom is cast aluminum, and the top is sheet metal. The front is a cast aluminum plate where the connectors are mounted. A unique aspect of this computer design is that it is completely sealed. The case components are assembled with a silicon gasket, and the connector bodies are sealed against the front plate. The assembly appears to be watertight.

The PCB is mounted to the connector plate and the bottom of the case. When assembled, the case contacts the PCB in order to become a heat sink for two internal rows of output drivers. The drivers are not mounted to the heat sink by the tab provided, but are held in place using a compressing wedge in the case. This wedge holds the components in place against the heat sink material without the need for fasteners.

Design

The circuit board utilizes surface mount technology for the integrated circuits and mostly passive components (resistors, coils, and capacitors). The output driver devices are larger than most, indicating higher power devices. There is a lot of circuitry in the 1999 computer, although the IC count is only 11, which is an average number. The 1997 computer has a lot less, with 10 ICs, and fewer drivers. The PCB is not conformal coated in either unit, relying on the sealed case for protection. If the unit is built in the open, even in a clean room, there will be moisture (humidity) sealed in the case. The case could have been sealed in an inert atmosphere, such as Nitrogen or Argon to provide a better internal atmosphere, but this is costly and difficult and probably not the case. Hermetically sealed cases are an elusive thing, and it takes a perfect seal to maintain a pure internal atmosphere.

Production Processes

Most devices, even hermetic military and space grade, will transfer gas and moisture in and out. This occurs with temperature cycles and time, sometimes at little more than the molecular level, and degrades the internal atmosphere. The exposed PCB would be susceptible to organic growth and corrosion over time. Insects probably cannot get in. There is an electrolytic capacitor on the PCB that is part of the power supply circuit. These capacitors are liquid filled, and can develop leaks with temperature cycles and time. The chemical can be very damaging to the circuitry, and has been known to eat through the conductors on the PCB. (Mitsubishi had considerable difficulty with this problem on computers built in the late 1980s and early 1990s, sending scores of their vehicles to the scrap yards by the end of the decade.)

Conformal coating would provide some protection against this as well as the other moisture related problems. The thick aluminum frame/bottom assembly is very rigid, with little potential for flexing or warping due to vibration or temperature cycles. The case is closed with standard Torx head fasteners. All the fastener holes in the case are tapped.

The PCB is double sided, with components on both the top and bottom of the PCB. All of the ICs are on the top side, so the density is only average. All of the ICs are surface mount, with leads, as opposed to leadless chip carriers (LCC) that are more subject to solder cracking under shock and vibration stress. The through-hole components are limited to the drivers and a few large passive devices. The integrated circuits are all in plastic packages and the size and pin counts indicate sophisticated devices. As with most of the other computers, some of the ICs have the marking removed. The soldering is uniform and well done.

There are four connectors, used as halves, to connect to the two connectors in the harness. They are high grade, high-reliability connectors with gold plated contacts. The connectors are sealed on the outside with a type of conformal coating or epoxy that should provide good corrosion resistance. The connector sockets and the wiring harness plugs are positively secured with threaded shafts. This prevents the connector from coming apart under shock and vibration conditions.

Comments

Construction-wise, this is a sturdy robust mechanical design. The lack of conformal coating is a reliability concern, despite the sealed case. The Z71 is an off-road package, and any truck can be reasonably expected to see harsh duty. The opportunity for moisture incursion will also be determined by the location of the ECU in the vehicle.

Fig. 5. **ECU for 1999 Chevy Z71**
The case components are assembled with a silicon gasket.

Fig. 6. ECU for 1999 Chevy Z71

There is an electrolytic capacitor on the circuit board that is part of the power supply.

Fig. 7. ECU for 1999 Chevy Z71

All of the integrated circuits for this module are surface mount.

Subject Computers: ECU for 1999 GMC Yukon; 1996 Chevrolet S-10 Pickup

Mechanical Aspects

The Yukon computer case is a clamshell design, using cast aluminum top and bottom halves. The case is coated with a black epoxy or paint on the outside. The two case halves have an o-ring seal between them, however the connector openings are not sealed, so the case is not watertight. There is an access cover, also with an o-ring, to allow access to a device that is mounted in a removable plastic carrier. The purpose of the o-rings cannot be to seal the case, but may be to prevent the generation of conductive aluminum particles from the case parts rubbing under vibration. The case is closed with standard Torx head fasteners. All the fastener holes in the case are tapped.

Design

The computer uses two PCBs arranged in a sandwich fashion. The circuit boards utilize surface mount technology for the integrated circuits and mostly passive components (resistors, coils, and capacitors) and through-hole for the drivers and a few large passive devices. Both PCBs are double sided, with components on both sides of the PCB. All of the ICs are surface mount, with leads, and are on the outside of the sandwich. The inside contains only passive devices. The integrated circuits are all in plastic packages and the size and pin counts indicate sophisticated devices. As with most of the other computers, some of the ICs have the marking removed. The output driver devices are larger than most, indicating higher power devices. The two PCBs are connected with a flat ribbon cable.

The case is finned on both halves, and is used as a heat sink for the drivers. Like the other Chevrolet computers, the drivers are not attached to the case. The design is different than the other computers as there is no wedge. The drivers are mounted upside down so that the tab contacts the case when the clamshell is closed.

Production Processes

This design requires critical dimension control of the PCB sandwich assembly to ensure contact with the case. The PCB sandwich is separated by rubber or plastic standoffs that allow the assembly to compress. A strip of this material runs the width of the PCB under the drivers. When the case is assembled, the halves compress the PCB about 2 millimeters, based on the gap in the case. The case supports the PCB assembly where it contacts the drivers and the corners on the far end. It cannot be determined if the sandwich assembly compresses uniformly or puts the boards in bending stress. The question here is what affect this bending stress will have on the reliability of the unit. The thick aluminum case halves do form a rigid assembly, however, any vibration the unit is exposed to will be transmitted to the PCB assembly through the driver—case interface.

The soldering is uniform and well done. The entire sandwich is conformal coated using a dip process as opposed to a spray. In order to keep the connector free of the conformal coat, the assembly is not fully submerged. This leaves a few conductors and a capacitor at the connector end uncoated.

The removable plastic carrier is the approximate size and shape to hold an EPROM; however, it actually contains a thickfilm hybrid (a multichip module). A hybrid is a super miniaturized circuit, and thickfilm is a ceramic substrate that is used instead of a PCB. Small and light, this technology is frequently used for complex circuits intended for flight. Most vehicle computers that use this technology use it for unsophisticated passive circuitry; however, this device does contain an IC. The marking on the IC is removed, but one can speculate that this is a programmable device. The entire hybrid is thickly coated for moisture protection. The module mounts by the 8-pin connector, and has locking tabs to secure it.

Comments

There are four connectors, and although three appear identical, they are keyed differently and cannot be cross-matched with the wrong plug. The case has the connector colors cast into the case in big letters, so hookup appears foolproof. The connectors are high grade, high-reliability connectors with gold plated contacts, and metal shells. The case has tabs for the harness plugs to lock on to, to keep them connected.

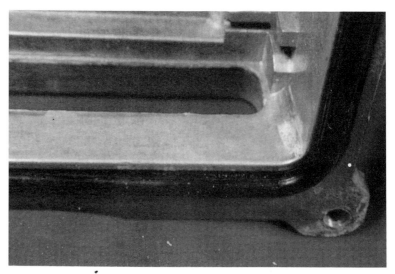

Fig. 8. **ECU for 1999 GMC Yukon**
The two case halves have an O-ring seal between them.

Fig. 9. ECU for 1999 GMC Yukon
Both circuit boards are double sided.

Fig. 10. ECU for 1999 GMC Yukon
The case is finned on both halves and is used as a heat sink for the drivers.

Fig. 11. **ECU for 1999 GMC Yukon**
This unit contains a thickfilm hybrid multichip module.

Fig. 12. **ECU for 1999 GMC Yukon**
The connector colors are cast into the case, making hookup foolproof.

Chevrolet S-10 Pickup

Mechanical Aspects

The S-10 computer case is a clamshell design, using cast aluminum top and bottom halves. The two-case halves have an o-ring seal between them; however, the connector openings are not sealed, so the case is not watertight. There is an access cover, also with an o-ring, to allow access to a hybrid device that is in a carrier (this carrier is identical to the Yukon computer). The cover has a cork block glued to it to prevent the hybrid from moving under shock and vibration conditions. There are no markings to indicate placement of the block, so it appears to be an afterthought, or perhaps a corrective action to a problem. The case is used as a heat sink for the output drivers, and this is accomplished in a conventional fashion. The drivers are mounted by their mounting tab to an aluminum bar that is bolted to the case.

Design

The computer uses two PCBs arranged in a sandwich fashion. Each of the PCBs is bolted to its respective case half. The board flex is controlled with dampers, in the form of rubber spacers in strategic places in the unit. The aluminum case halves form a rigid assembly, however, any vibration the unit is exposed to will be transmitted to the PCB assembly through the driver-case interface. The case is closed with standard hex-head fasteners. All the fastener holes in the case are tapped.

The circuit boards utilize surface mount technology for the integrated circuits and mostly passive components (resistors, coils, and capacitors). For the output drivers and a few large passive devices, through-hole parts are used. Both PCBs are double sided, with components on both sides of the PCB. All of the ICs are surface mount, with leads, and are located on the outside surfaces of the sandwich. The inside contains only passive devices. The integrated circuits are all in plastic packages and the size and pin counts indicate sophisticated devices. As with most of the other computers, some of the ICs have the marking removed. The two PCBs are connected with two flat ribbon cables.

Production Processes

The soldering is uniform and well done. The entire sandwich is conformal coated as a unit. The coating is such that it could be either a dip process, or a mask and spray process. Either way, there is about a centimeter of the boards uncoated, but there are no components and few conductors in this area.

The hybrid appears identical to the Yukon hybrid. The removable plastic carrier is the same approximate size and shape and contains a thickfilm hybrid. This device also contains a single IC. The marking on the IC is removed, but is a probably a programmable device. The entire hybrid is thickly coated for moisture protection.

Comments

There are three connectors, similar in construction to the Yukon connectors, except they have a plastic shell. They are keyed differently and color-coded to assure correct hookup. The connectors have gold plated contacts. The connectors are a commercial, high reliability grade. The case has tabs for the harness plugs to lock on to, to keep them connected.

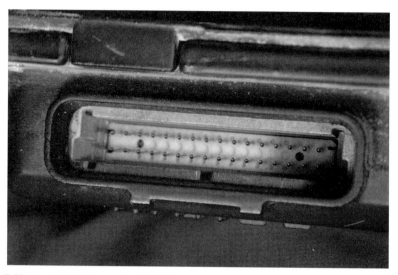

Fig. 13. **ECU for 1996 Chevy S-10 Pickup**
These connectors have a plastic shell.

Subject Computer: ECU for 1994 Chevrolet S-10 Pickup, 4.3 Liter V6

Mechanical Aspects

This computer uses a cast aluminum "bathtub" type case. The top is a flat aluminum lid. The lid is very heavy, almost 3mm thick. The case is coated with a black epoxy or paint on the outside. The case and the lid have an o-ring seal between them, however the connector openings are not sealed, so the case is not watertight. There is an access cover, also with an o-ring, to allow access to a device that is mounted in a removable carrier. The case is used as a heat sink for the output drivers, with the drivers mounted by their mounting tab to an aluminum bar that bolts to the case. The case is closed with standard Torx/hex- head fasteners. All the fastener holes in the case are tapped.

Design

The computer uses a single PCB that is well supported around its perimeter. The circuit board has a similar construction than the other GM computers, with surface mount technology for the integrated circuits and mostly passive components (resistors, coils, and capacitors). The output drivers and the larger passive devices are through-hole parts. The PCB is double sided, with components on both sides of the PCB. All of the ICs are surface mount with leads, and are on one side of the PCB. The integrated circuits are all in plastic packages. The size and pin counts indicate sophisticated devices. Unlike most of the computers, all of the ICs have the marking intact. The PCB is fully assembled and then conformal coated, including the heatsink, using a mask and spray process. The soldering is uniform and well done.

There is an unusual aspect to this design. There is a thick braided wire running the length of the heat sink assembly, placed so it will scratch against the case. This would probably provide a positive electrical connection to the case, since the heat sink assembly is coated with the PCB and would be insulated everywhere else. This has the appearance of an afterthought, and may have been a corrective action to solve a problem. The thickness of the braid requires that there was room allowed for it in the design, so it may have been originally designed in. There is no way to know which. The reliability concern is that it will generate conductive particles under vibration. Since the board is well coated, and has no exposed circuitry, this may not be a problem.

Production Processes

The removable carrier is the approximate size and shape of a small Dual Inline Package (DIP) microprocessor and is made out of plastic. It contains a sandwich module with a thickfilm hybrid and a small PCB. The PCB holds the connector pins and two DIP ICs. The marking on the ICs is not accessible without destroying the device. The hybrid is thickly coated on the component side;

however, the underside and the PCB are not coated. This is a reliability concern. The module mounts with a 36-pin connector with locking tabs.

Comments

There are four connectors. Three are conventional computer connectors, and one is a power connector with high current pins. The three are similar, and

Fig. 14. **ECU for 1994 Chevy S-10 Pickup**
The case and the lid have an O-ring seal.

Fig. 15. **ECU for 1994 Chevy S-10 Pickup**
The computer uses a single circuit board that is well supported around its perimeter.

uniquely keyed to prevent cross connection. As with other GM computers, the case has the connector colors cast into the case in big letters. The connectors have gold plated contacts, and metal shells. These are high grade, high-reliability connectors. The case has tabs for the harness plugs to lock on to, to keep them connected.

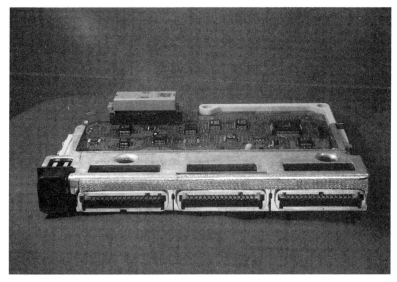

Fig. 16. **ECU for 1994 Chevy S-10 Pickup**
This unit has a thick braided wire running the length of the heat sink assembly.

Fig. 17. **ECU for 1994 Chevy S-10 Pickup**
This unit contains a sandwich module with a thickfilm hybrid and a small circuit board.

Subject Computers: ECU for 1993 Mazda Navajo 4X4, 1995 Ford Explorer 4X4

Mechanical Aspects

The 1993 and 1995 computers are similar, both utilizing a cast aluminum case frame with a sheet metal top and bottom. The aluminum case is used as a heatsink for the output drivers and they are attached to the side of the case with clips. The assembly is not sealed. The circuit board is conformal coated for protection against foreign material, moisture, and organic growth, such as fungus or insects. This provides a sturdy assembly with little potential for flexing or warping due to vibration or temperature cycles. The case is closed with Torx fasteners, probably more for ease of assembly than tamper resistance. The single circuit board is securely fastened to the aluminum frame with 6 Torx screws, one of which has been filled with solder (reason unknown).

Design

The circuit board utilizes both through-hole and surface mount technology. All of the ICs are through-hole type. Through-hole technology is an established technology dating back to the beginnings of PCB design. It is very rugged, and is more tolerant of solder degradation, as the component leads are soldered on both sides of the PCB. The ICs are all plastic packages, except for the EPROM, which is in a ceramic package. Ceramic packages are used for increased operating temperature range. The only surface mount components are resistors and capacitors.

Production Processes

There is a single connector with tin-plated contacts. The tin-alloy pins (male) appear to be standard commercial or high-reliability grade (military grade and above have gold plated connector pins). The two-connector halves are positively secured with a threaded shaft. This prevents the connector from coming apart under shock and vibration conditions.

Comments

There is an internal "edge" connector on the PCB, which has been rendered inoperable by the application of an insulating coating, separate from the conformal coat. This connector is probably used to program or test the unit during assembly. There is a covered opening in the case, suggesting the need for access after unit assembly. This may be due to the programming being done after complete assembly, or the need to extract information at a later date.

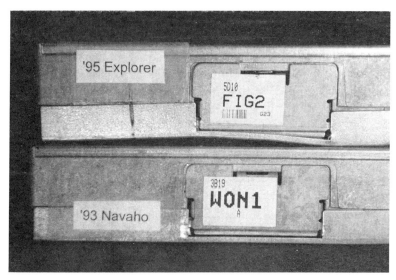

Fig. 18. ECU for 1993 Mazda Navajo 4X4
The aluminum case is used as a heatsink for the output drivers.

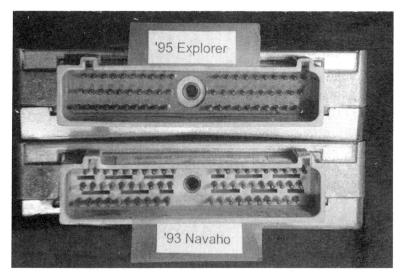

Fig. 19. ECU for 1995 Ford Explorer 4X4
This is a single connector with tin-plated contacts.

Subject Computer: ECU for 1999 Ford Explorer 2X4

Mechanical Aspects

The 1999 computer utilizes similar construction to earlier Ford models, with a cast aluminum case frame with separate top and bottom. The top is a simple sheet metal top. The bottom, however, is cast aluminum. This is done to facilitate easier PCB mounting at the edge of the frame. The bottom becomes part of the frame when attached, and provides clearance for the PCB and components. The case is used as a heatsink for the output drivers in the unit, however, in an internal fashion. The frame extends under the PCB along the sides, which allows for heat to dissipate from the components without their being clipped to the sides of the frame. The mass of the cast aluminum bottom also adds to the heat dissipation of the frame. This allows the PCB to be fully assembled, including the drivers, and attached to the frame as a single assembly.

Design

The circuit board utilizes modern surface mount technology and is assembled utilizing automated component placement (pick and place technology). This PCB is double sided, with components, including integrated circuits, on both the top and bottom of the PCB. This allows for a denser population, i.e. more circuitry per square inch. The soldering is uniform and well done, and appears to be totally automated, such as a utilizing solder paste and a reflow oven. This allows for precise process control of the solder process, and produces uniform solder joints. Eliminating the variables from hand soldering, such as iron temperature, iron contact duration, and solder quantity, increases reliability. All of the ICs are surface mount, with leads, as opposed to leadless chip carriers (LCC) that are more subject to solder cracking under shock and vibration stress.

The integrated circuits are all in plastic packages. The size and pin counts indicate sophisticated devices. Most of the ICs have the marking removed, perhaps on purpose for design protection, but possibly due to the cleaning processes associated with automated PCB assembly. The latter, however, would probably remove all of the marking from all the devices, which is not the case.

The computer is not sealed, however the circuit board is conformal coated to protect against foreign material, moisture, and organic growth. The thick aluminum frame and bottom assembly is very rigid, with little potential for flexing or warping due to vibration or temperature cycles. The case is closed with standard hex-head fasteners, with tapered threads for ease of assembly. The receiving holes are tapped, eliminating the generation of conductive aluminum shavings or particles associated with self-tapping hardware. The single circuit board is securely fastened to the aluminum frame with six hex screws, two of which have been covered with solder. It was noted that the PCB is assembled to the case frame prior to conformal coating. While this was probably done for efficiency reasons, it does seal the PCB to the frame along the edge, and locks the PCB attachment screws in place. This would provide some locking action, which might prevent the screws from coming loose under rough conditions.

Production Processes

The connector used has gold plated contacts, which indicate a high-reliability grade or better connector, with good corrosion resistance. The connector halves are positively secured with a threaded shaft to prevent the connector from coming apart under shock and vibration conditions.

Comments

Construction wise, this represents the current high quality automated assembly process. There are 8 obvious output drivers, with considerable circuitry to control them. This unit appears to have more computing power than the 1994–1995 modules. There is an internal "edge" connector on the PCB, which has been rendered inoperable by the conformal coating. This connector is probably used to program the unit during assembly; however there is a covered opening in the case, suggesting the need for access after unit assembly. This may be due to the programming being done after complete assembly, or the need to extract information at a later date.

Fig. 20. **ECU for 1999 Ford Explorer 2X4**
The frame extends under the circuit board, allowing for heat to dissipate.

Fig. 21. ECU for 1999 Ford Explorer 2X4
The computer is not sealed, but the board is coated for protection.

Fig. 22. ECU for 1999 Ford Explorer 2X4
There is an internal "edge" connector that was rendered inoperable by conformal coating.

Subject Computers: ECU for 1988, 1998 Toyota Pickups

Mechanical Aspects

This unit is not an ECU, but is an ABS module. This unit uses a drawn sheet metal case closed with bend-over tabs. The PCB is 100 percent through-hole technology, with a single microprocessor, and little or no memory. The construction indicates a design much older than 1990, although no date coding was identified (the date may be the date pulled from the vehicle). The construction is 1980s technology, with a commercial quality level.

Design

The computers from the Toyota vehicles represent a 10-year span in technology. The 1988 computer is single sided, with all through-hole technology, and the 1998 computer is double sided, with mostly surface mount technology, and a few through-hole parts. Both boards are conformal coated. The units are similar in construction, with cast aluminum cases and sheet metal tops. The 1988 unit has a separate sheet metal bottom, while the 1998 unit has an aluminum bottom that is part of the frame.

The output drivers in both computers are attached to separate heat sinks, which attach to the frame. The connectors are plastic bodied with tin connectors. The computers are similar in construction as the Nissan ECUs, indicating an adherence to the same manufacturing philosophy. These units are simply current commercial practice for the time they were built.

Comments

Due to the similarity of construction between the Nissan and Toyota computers, I believe this represents a fundamental difference from the American manufacturers approach to the ECU. The Japanese ECUs appear to be a "best commercial practice" approach (with allowances for the application) as opposed to the custom approach of the American computers. Although the Toyota computers show no manufacturer's name, they appear to be a sub-assembly that possibly is contracted out to companies that specialize in such assembly and technology. Utilizing a company that is accustomed to building a specific type of product (electronic as opposed to automotive) will inherently increase the available technology and experience applied to the design and construction of the computer. This, in turn, can be said to increase reliability, as lessons learned can be applied on an ongoing basis, and do not have to be re-learned.

Subject Computers: ECU for 1988, 1991, and 1996 Nissan Pickups

Mechanical Apsects

The Nissan computers are grouped together because they are all very similar. The basic structure is unchanged over the three computers, which range from 1988 to 1996. A computer not shown here, from the 1984 300ZX, is made the same. This leads to the conclusion that Nissan uses a more standardized approach to computer modules.

Design

The computers are housed in an aluminum frame with sheet metal top and bottom. Allowances are made in the aluminum frame for the different PCBs, but the basic design is the same. The 1988 computer is very conventional for the time frame, with a commercial microprocessor and EPROM memory. All of the components are through-hole. The PCB is single sided, with all the components on one side. The 1991 computer utilizes surface mount ICs and a mix of surface mount and through-hole passive components. The 1996 computer has a similar PCB, although with fewer through-hole components. The 1991 and 1996 computers have components on both sides. All of the output drivers are mounted to the case frame on all three computers. None of the cases are sealed; however, all the PCBs are conformal coated.

The connectors are also similar, with plastic bodies and tin plated pins.

Production Processes

The assembly design of the computers is interesting in that all of the mounting structure is attached to the bottom cover. This allows the use of the same computer in different vehicle mounting scenarios just by changing the bottom cover. All of the computers are manufactured by Hitachi, which may account for their similarity.

The American manufacturers' modules appear, for the most part, to be individually designed for a specific vehicle. The Nissan computers reflect a manufacturing philosophy of standardization, where only needed items are changed. This would infer an adherence to the thought that the less is changed, the less impact on reliability. The similarity among the units may also indicate a more utilitarian approach to the ECU as a disposable module. Although these units are as reworkable as any, it is a reality that few facilities exist for repairing ECUs, and most will be swapped out when they fail.

Comments

The simpler approach to design will keep the unit cost low, and lessen the impact to the consumer who has the misfortune of having an ECU failure. This is not to say that the modules are cheap, as the technology represents the standard best commercial practice of that time. Taking a good design and modifying it as you go is almost always preferable to a complete redesign, from a cost and reliability standpoint.

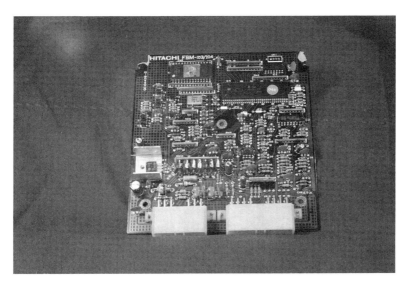

Fig. 23. **ECU for 1988 Nissan Pickup**
All of the components for this module are through-hole.

Fig. 24. **ECU for 1996 Nissan Pickup**
This module has relatively few through-hole components.

Fig. 25. **ECU for 1996 Nissan Pickup**
This module has relatively few through-hole components.

Subject Computers: ECU for 1990 Isuzu 2.3 Liter Pickup, 1990 Isuzu Trooper

Comments

This unit does not appear to be an ECU, based on the circuitry, as compared to the other 1990 computer modules. The unit is housed in a plastic case that is held closed with snaps at each corner, and a single screw in the center of the unit. The PCB is 100 percent through-hole technology, with only three ICs and three drivers. The board is conformal coated. The connector is plastic, with 17 tin plated pins. If this is an ECU, it is very basic, and cannot control very much. The circuitry is early 1980s technology, with a commercial quality level.

1990 Isuzu Pickup

Mechanical Design

This computer was built by Hitachi. The case is formed sheet metal, and closed with bend-over tabs. No attempt was made to keep out moisture, or dirt for that matter.

Design

The PCB is a relatively simple single sided board with through-hole technology. Only one of the output drivers is mounted to a heat sink, and it is not tied to the case. The other 7 drivers are in the air, and this is a reliability concern. The PCB is conformal coated, but only lightly so. The connector end was taped off leaving an inch of exposed circuit board. This area contains a few passive components.

Production Processes

There is also a coating void on the circuit board. Although no components are exposed, this is definitely a process problem, and probably an inspection escape. There is an EPROM in the circuitry. This assembly may have had the components mounted by hand, but it was soldered using an automated process.

Comments

The soldering quality looks good. The unit connects with two separate connectors that have tin plated pins and no apparent means of locking the plugs in. This unit represents early 1980s technology and a bare bones approach to the unit. It is only commercial quality at best.

Fig. 26. **ECU for 1990 Isuzu Pickup**
This is a straightforward design using through-hole technology.

1990 Isuzu Trooper

Mechanical Aspects

This unit bears a close resemblance to the 1989 S-10 computer. The case is identical except for the location of the access hole. Several of the large ICs are Delco parts. There is no doubt that this is a GM computer

The unit is housed in a stamped and folded sheet metal case. The case has no moisture protection, and has small openings at the corners that dirt and insects can get through. There is an access cover, with a gasket, to allow access to a device that is mounted in a removable carrier. The case is closed with hex-head fasteners with tapered ends. These fasteners could tap their own threads; however, the condition of the metal around the holes indicates they were tapped.

Design

The computer uses a dual PCB design that is arranged in sandwich fashion, however, the top board is more of a daughter board, only a couple of inches wide. It holds the programmable devices and has an edge connector for programming. There is clearance for this connector in the case, indicating a need to connect after assembly. The two boards are connected with a ribbon cable. The boards are assembled together with a metal framework and installed as a unit. Part of this framework provides heat sinking for the output drivers. This part of the framework bolts to the case. There is a second driver attached to a free standing heat sink, and this also presses against the lid. The PCB assembly is conformal coated as a unit, including the edge connector.

Production Processes

The main circuit board is fairly dense, and is all through-hole technology. The PCB is single sided, and has more than the usual number of large ICs. The daughter board holds only two ICs and they are housed in carriers. They are not readily removable.

Comment

There is a single ganged connector that accepts two plugs from the harness. The body is plastic and the pins are tin. The connector shell has locking tabs. This is a standard commercial grade connector. Although it appears cheaply made because of the case, it was built with good commercial practices.

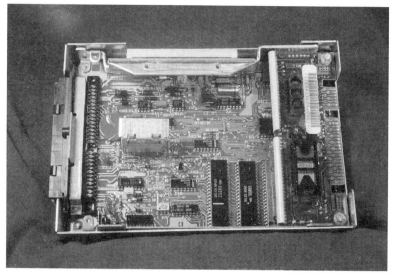

Fig. 27. **ECU for 1990 Isuzu Trooper**
This computer uses a dual circuit board design arranged in sandwich fashion.

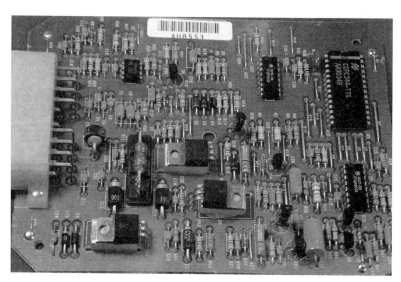

Fig. 28. **ECU for 1990 Isuzu Trooper**
This circuit board is fairly dense and is all through-hole technology.

Subject Computer: ECU for 1996 Dodge Dakota V8

Mechanical Aspects

This computer is housed in a bathtub style molded plastic case, with a molded lid. The lid has a raised section that is louvered for airflow. The PCB sits in the bottom of the case, and is potted with a flexible potting compound.

Design

The drivers are attached to heavy, finned heat sinks that extend out of the potting into the louvered area of the case. The potting is semi-transparent, but not enough to describe the circuitry. The connector has gold pins, and shows evidence of connector grease, which is used to keep moisture out of the connector. The connector is keyed, and accepts a locking screw from the harness plug. The case is closed with Torx head, self-tapping screws.

Comments

The self-tapping screws are of no consequence, even though some black plastics are conductive, because the circuitry is completely encased.

Fig. 29. ECU for 1996 Dodge Dakota
This lid has a raised section that is louvered for airflow.

Fig. 30. **ECU for 1996 Dodge Dakota**
This unit shows evidence of connector grease, which is used to keep moisture out of the connector.

Subject Computer: ECU for 1999 Dodge C-1500

Mechanical Aspects

This computer is housed in a bathtub style cast aluminum case, with a sheet metal lid. The PCB sits in the bottom of the case, and is potted with a flexible material. The drivers are attached to the side of the case for heat sinking. The potting is fairly transparent, enough so to see the circuitry.

Design

The PCB assembly is almost completely surface mount devices, with only a few through hole devices The connector is a single piece, ganged together to accept three harness plugs. The socket sections appear very similar, but they are keyed to accept only the correct plug. The connector body is plastic and has locking tabs. The pins are gold plated.

Fig. 31. **ECU for 1999 Dodge C-1500**
This unit is potted with a flexible material.

Subject Computer: ECU for 1998 Dodge C-1500

Mechanical Aspects

The 1998 computer is similar in appearance to the 1999 computer, although the case is formed sheet metal, and the lid is split at the connector exit. One oddity is that the lid is secured with bend-over tabs around the edges, with two screws inboard, next to the connector. The internal construction is very unique. As with the other Dodge computers, this one is potted. The large section of the lid is actually a second PCB, which is also potted using a foam rubber dam. This potting material is a soft gel, as opposed to the firmer potting material used in the main case. The two screws secure the connector that connects the main PCB to the lid circuitry.

Design

The PCB is almost completely surface mount devices, with only a few through-hole devices.

The connector is a single piece, ganged together to accept three harness plugs. The socket sections appear very similar, but are keyed to accept only the correct plug. The connector body is plastic and has locking tabs. The pins are gold plated.

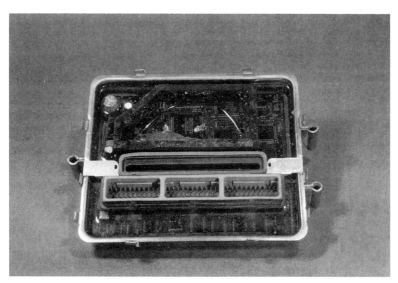

Fig. 32. **ECU for 1998 Dodge C-1500**
Like the other Dodge computers, this one is potted.

Production Processes

Devices are potted for many reasons. The potting is most effective at keeping out contaminants, dirt, and moisture. The PCB and components essentially become a solid block, so vibration is no longer an issue. There is also a reduced susceptibility to shock damage. Potting also prevents any rework after the potting operation, making failed units unrepairable.

There is additionally the issue of added security, as a lot of solvent deprocessing would be required to expose the PCB for analysis.

Comments

Most companies that pot devices to prevent analysis will use opaque material, or will remove the markings from the devices. One of the computers had the marking intact. I believe the decision to pot the computers was made to increase the ruggedness and reliability of the unit.

4. New Vehicle Electronics Design Trends and Issues

Vehicle manufacturers in the past were among the most conservative in adopting non-core electronics. For example, features such as alarms and high-quality audio systems were available for over a decade as after-market items before auto makers began including them as standard equipment on most models. While this seemed incomprehensible from the standpoint of car and electronic enthusiasts, it did allow vehicle companies to take advantage of user acceptance and more mature feature designs.

Increased competition and more fickle user tastes have led to faster adoption of non-core features, as well as acceleration of advanced development of core items such as engine and transmission controls. Competition has also narrowed or eliminated per-vehicle profit margins, leading designers to search for innovative cost controls. Finally, weight reduction affects every design decision.

Since World War II, most vehicles have been built with 12-volt systems, an enduring standard, though some older readers will recall Volkswagen Beetles, Jeeps, and many other cars with six-volt systems. A significant change that will transform how vehicle electronics systems are designed is the move to 48-volt power.

The reasoning behind the move is simple: since higher-voltage systems can get by with lower current and still deliver the same power to components, a car built around a 48-volt system can use smaller (lighter) wiring.

Discussion

As noted above, virtually every aspect of modern vehicle performance is directed by sophisticated electronic sensor-analysis-control-correct systems. One obvious set of controls involves the engine. The air-fuel mixture, the ignition timing, and other functions are controlled to produce the greatest power and fuel economy. Another obvious control simply advances the automatic transmission as the vehicle needs greater power or speed. Still another puts vehicles such as the Explorer into four-wheel drive when the rear wheels slip.

Less obvious are controls designed to make large, awkward vehicles such as sport utility vehicles (SUVs) easier to drive. Some SUV brakes, rather than simply stopping the vehicle using the effective anti-lock system, also are engaged to assist in turning. The power steering hydraulic pressure is increased when the system senses it is encountering resistance. The engine, transmission, brakes, and suspension are worked together to give the truck-like SUV an easier ride in order to increase sales to persons not used to driving such vehicles.

All of these functions require computers contained in modules of varying complexity. These sophisticated systems contain state-of-the-art microprocessors and related memory that dealers often reprogram with the latest software designed to increase vehicle performance.

The result is that analysis of vehicle failure today is more akin to analyzing an airliner crash than in the days of Pintos, Edsels, or Corvairs. A computer malfunction could cause an engine to race out of control or brakes to engage or disengage completely without regard to driver commands. Automakers, obviously, take great care to build systems that are as fault-proof as possible and some very advanced electronics work is done in the automotive field. But when electronic design flaws occur, they are often difficult to detect, especially when they occur sporadically.

Investigations into electronic vehicle malfunctions must therefore determine:
- The extent to which the system's basic design and manufacturing make it susceptible to failure.
- The extent to which failures of all types have been documented in the vehicle.
- The repairs/recalls that have been made as a result of failures, both disclosed and undisclosed.
- The extent to which the subject system has been partly or completely redesigned in apparent recognition of design flaws.

Background

For this inquiry I reviewed literally thousands of pages of automaker and after market technical data, and module and sensors from a variety of vehicles. To provide the most interesting and informative discuss, once again I return to the highly innovative Ford Explorer electronics package as an example.

On Ford's Control Trac four-wheel drive system, the driver can select three different modes of operation including automatic four-wheel drive. The electronically shifting transfer case consists of front and rear Hall Effect-style drive shaft speed sensors, an electric four-wheel drive shift motor, and an electric clutch. The electric clutch is used to engage the front drive shaft to allow shifts into four-wheel drive at any speed—even when the front wheels are stopped, and when the vehicle has become stuck while still in two-wheel drive.

When in automatic four-wheel drive, the generic electronic module (GEM) controls power to the transfer case electric clutch in response to wheel slippage, as seen through the front- and rear-drive shaft speed sensors. When in low range mode, the electric clutch is fully engaged. In 1997, all Ford F series, F350-F450, Rangers and Explorers went with the GEM module to help with the electronic 4x4 systems. Ford's ESOF (Electronic Shift on the Fly) system relies heavily on the GEM for proper operation.

The Electronic Shift on the Fly (ESOF) system is an electronic shift 4x4 system that allows the driver to choose between two different 4x4 modes, as well as 2-wheel drive. The driver can switch between two-wheel drive and four-wheel drive high (4H) mode at speeds up to 55 mph (89 km/h). To engage or disengage the four-wheel drive low (4L) mode, the vehicle speed has to be below three mph (five km/h), the brake pedal applied, and the transmission in neutral. The transfer case is equipped with an electromagnetic clutch inside the case. This clutch is used to spin up the front driveline when shifting from two-wheel drive to 4H mode at road speeds. When you turn on the control switch, the instrument panel to the 4H mode, the GEM recognizes that a shift has been requested. It then activates the electromagnetic clutch and the relays that power the transfer case shift motor.

When the shift motor reaches the desired position (as determined by the contact plate position signal), the GEM removes power from the shift relays and motors. When the transfer case front and rear output shafts are synchronized, the spring-loaded lock-up collar mechanically engages the mainshaft hub to the drive sprocket. Finally, the front axle collar engages, and the GEM deactivates the electromagnetic clutch.

Significant Research Findings

One significant data point involves GEMs from wrecked 1997 and 1998 Explorers. These GEMs are based on a Motorola-designed board that contains a Texas Instruments controller/microprocessor. Motorola, obviously one of the world's largest designers and producers of such controllers, would not normally use one designed and built by a competitor. The most plausible explanation is that the design originally contained a Motorola processor and that Ford replaced it with a TI device after it discovered possible systemic failures.

A second significant data point concerns the design of the currently used Ford Explorer GEM. The older GEMs were enclosed in a loosely sealed plastic box that allowed the entry of not only moisture, dirt, and other contaminants, but also electromagnetic interference. The new GEM is sealed in a metal package that provides significantly more protection against all of these factors. The new GEM has a much higher inherent cost of manufacture, which implies that Ford found it necessary to increase its GEM production costs to correct failures.

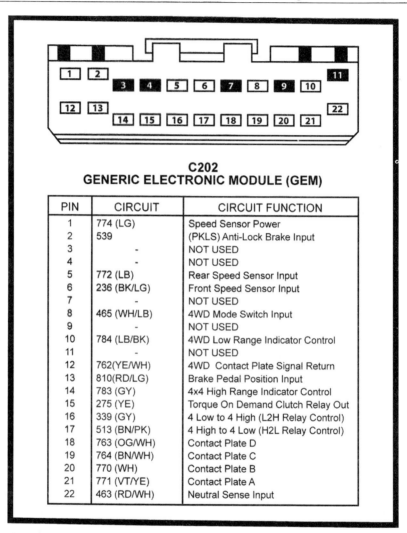

Fig. 33. **GEM Functions**

A third significant data point concerns the recall of 1999 and 2000 Ford Explorers because the generic electronic module (GEM) could experience a condition Ford refers to as "lock-up." This condition is known among semiconductor designers as "latch-up." The phenomenon is discussed in detail below. The "fix" employed in this recall is highly important. To address this problem, Ford mechanics were instructed to add a resistor to the GEM. Please note that this would not prevent the unit from latching up, but rather let the excess power drain from it after latch-up. Thus, the fix would not prevent a GEM-induced failure, and after a mishap there would be no trace that this had occurred.

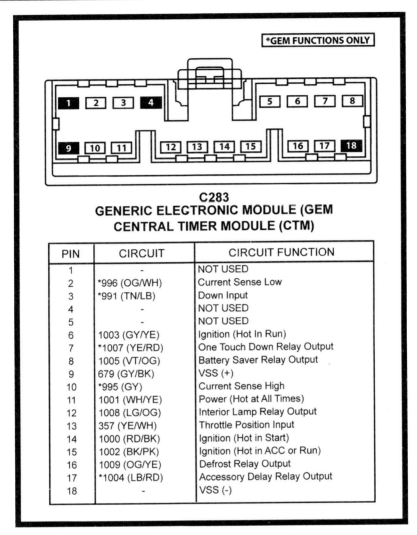

Fig. 34. GEM Circuit Functions

PIN	CIRCUIT	CIRCUIT FUNCTION
1	-	NOT USED
2	*996 (OG/WH)	Current Sense Low
3	*991 (TN/LB)	Down Input
4	-	NOT USED
5	-	NOT USED
6	1003 (GY/YE)	Ignition (Hot In Run)
7	*1007 (YE/RD)	One Touch Down Relay Output
8	1005 (VT/OG)	Battery Saver Relay Output
9	679 (GY/BK)	VSS (+)
10	*995 (GY)	Current Sense High
11	1001 (WH/YE)	Power (Hot at All Times)
12	1008 (LG/OG)	Interior Lamp Relay Output
13	357 (YE/WH)	Throttle Position Input
14	1000 (RD/BK)	Ignition (Hot in Start)
15	1002 (BK/PK)	Ignition (Hot in ACC or Run)
16	1009 (OG/YE)	Defrost Relay Output
17	*1004 (LB/RD)	Accessory Delay Relay Output
18	-	VSS (-)

C283
GENERIC ELECTRONIC MODULE (GEM
CENTRAL TIMER MODULE (CTM)

*GEM FUNCTIONS ONLY

Analysis of Explorer Semiconductor Failure

Defining the primary failure mode of CMOS ICs shows us a method of how to protect modern COTS ICs. We know that the primary failure mechanism of all modern systems that use CMOS (latch-up) ICs is a much more sensitive failure mechanism than physical failure. We know how to test any IC as to its particular sensitivity by injecting DC currents into interface pins and we can also expose equipment to realistic levels of RF signals and then measure the equipment response.

Because there are two primary designs of the Electronic Control System there must be two analyses. The first design covered the period of 1991-1994, when an

OBD-1 system was used. This system utilized the General Electronics Module, a Power Train Control Module, as well as additional modules for the Automatic Braking System, the Electronic Control System, and the "Transmission Controller."

The second model of the electronic system started in 1995 and was called the OBD-2 system. In this system many of the functions executed by external modules in the OBD-1 system were now incorporated into the Power Train Control Module. This system also incorporated a separate General Electronics Module and Electronic Control Module. However, the Transmission Control was now incorporated into the Power Train Control Module and significant changes were made in the system communications bus, which now consists of a two wire serial bus that has a self-correcting feature which allows self-modification of transmitted data in the event that the data is corrupted during transmission.

It is the second version of the system that was recalled in 1999 and 2000[4], which had a problem with the GEM unit locking up, causing the windshield wipers to come on, and disabling the dome light and the driver's window control, as well as the ability to enter four-wheel drive mode.

This failure was also identified by a trouble code, which was displayed on the OD (probably overdrive) light on the dashboard. From the description of the failure symptoms it is clear that a latch-up problem in the GEM module has an effect on the function of the Powertrain Control Module (PCM) in that it prevents proper operation of the transmission control.

It is possible to construct a scenario whereby the normal safety features of this system, which are primarily constructed from vehicle speed and transmission speed sensors, are duplicated. These safety features could be bypassed in the event that the PCM unit were subjected to a latch-up condition, which could cause a low range four-wheel drive control signal to be issued even though the vehicle speed sensor was indicating that the vehicle was moving at a high speed.

In addition to the potential failure of the Transmission Control system, the automatic braking system and the traction control system, which are also controlled by the PCM, would also be susceptible to failure. It is my opinion that the inappropriate issuing of commands of these systems could cause a brake to lock-up, perhaps on only one wheel, possibly the left rear wheel, which could result in damage to a tire and subsequent lack of driver control, resulting in a rollover.

---

[4] NHTSA Recall No. 00V072/Ford Recall No. 00S04. *Ford Motor Company.* Models: Ford Explorer (1999-2000), Mercury Mountaineer (1999-2000). Number Involved: 208,903. Dates of Manufacture: September 1998 - September 1999. Defect: Sport utility vehicles equipped with 4.0L engines and All-time 4-Wheel Drive (A4WD) powertrains. The generic electronic module (GEM) could experience a condition referred to as "lock-up" in which the GEM controlled electronic functions (e.g., front windshield wipers, interior lights, 4x4 system, etc.) could not be turned ON or, in some cases if the function is ON, could not be turned OFF. Remedy: Dealers will install a resistor in the GEM circuit. The manufacturer has reported that owner notification began April 15, 2000. Owners who do not receive the free remedy within a reasonable time should contact Ford at 1-800-392-3673.

5. Discussion of Failure Mechanisms, Including Case Studies

Automotive manufacturers possess a great deal of test and service data. Analysis of this data would let them know quickly that such failures were occurring. But this considerable test and service information, is very hard to detect and document from the outside. The clues provided to outside investigators are subtle. This book draws on no special relationships or conversations with any vehicle manufacturer, but rather is based on information available to the general public. This involves reviewing technical data supplied to repair facilities and published consumer complaints, plus an independent examination of systems of interest.

Having selected the Explorer as the primary case study, I researched brochures going back to the vehicle's introduction in 1991. I discovered that advanced electronic systems and options have always been a focal point of marketing the Explorer line. A key feature has been the ability to switch electronically between two-wheel drive (2WD) and four wheel drive (4WD) by the simple flick of a switch. This is accomplished by using a specially designed Borg-Warner electronic transfer case. The later models of the Ford Bronco also shared this feature; however, virtually all other four-wheel drive vehicles at this time used a manually shifted, floor-lever system. Though other SUVs have since adopted this electronic transfer case shifting technology, the Explorer system led the way by many years.

The importance of the Explorer introducing this electronic four-wheel drive system from a marketing perspective cannot be overstated. Many vehicle electronic systems, such as the seat belt interlock chip I built for Ford, were regarded by automakers and nearly all consumers as government-imposed nuisances. The nuisance factor far overshadowed the fact that these electronic systems replaced mechanical ones. Others, like a built-in citizens band (CB) radio I built for Cadillac were fads, novelty "toys" installed for only a short time to entice the consumer.

The development of small, sophisticated, computer controlled systems used to automate mechanical systems significantly expanded the range of features that could be offered to attract new market segments. The electronic transfer case was a core part of the Explorer systems that greatly expanded the vehicle's commercial appeal. Prior to this innovation, four-wheel drive vehicles required manual control of the hubs and drive system, and drove more like tractors than cars. "Soccer moms," who drove minivans and before that station wagons, now had the option of the more appealing SUV. Ford created an SUV that all the drivers in a family could operate easily and comfortably.

The electronics development and integration task faced by Ford's engineers was, to use a Silicon Valley cliché, "non-trivial." The deadlines and cost pressures I faced when designing chips for military systems were mild compared with those imposed on the Explorer design engineers.

To understand the scope of the project, imagine for a moment developing capabilities that will be used by a wide variety of the population, very few of whom will read the instructions that you provide to them. The systems must be simple to operate and "idiot-proof" while at the same time doing everything the marketing department has promised consumers.

The components have to survive in a very harsh environment, yet the system must operate for years without failure so that warranty costs do not eat up initial profits. Every penny of cost and every gram of weight possible must be squeezed out of the system. Further, the system will change virtually every year in response to marketing innovations, consumer complaints, field failures, cost-cutting moves, suppliers entering or exiting the business, and government regulations. It is against this backdrop that the Explorer's design must be viewed.

Ford's Brain: The Generic Electronic Module (GEM)

The brain, heart, and soul of the Explorer's electronics is the Generic Electronic Module, or GEM, which was introduced in 1994. "Generic" is a bit of a misnomer—I at first thought it might be a module used in a wide variety of vehicles. I found instead that it is a class of modules that consolidate a wide variety of functions in only a few vehicle families.

Ford's GEM is an attempt to combine a wide variety of sensor-driven electronic controls into one box and cut costs by eliminating components. In many vehicles, small microprocessors called controller chips and other integrated circuits are distributed and wired so that each handles a separate function such as seat belt engagement detection or anti-lock braking. In a fierce cost-cutting environment such as that imposed by former Ford CEO Jacques Nasser, a $15 dollar controller chip is a tempting target for designers, and each non-GEM vehicle had plenty of them.

The GEM controls the electronic transfer case function described above and also—in a design decision that receives significant attention in this book—includes controls for other simple functions such as windshield wiper operation. This requires the GEM's lone processor to handle many tasks at the same time rather than focusing on the critical task of four-wheel drive control. This opened the door for new types of failures to occur. First, the wiring required for these other functions creates paths for high-voltage currents to surge into the processor, causing it to latch up. Second, the processor might become overloaded with errant signals from these non-critical functions resulting in, for example, four wheel drive engagement/disengagement mistakes.

The decision to remove the mechanical four-wheel drive engagement lever as a failsafe created a situation where the electronics system must operate without mishap. Ford's implementation of an electronic transfer case system with high-risk electronics was a significant challenge prone to error. An electronic transfer case control is a design problem that should have elicited a high-end, reliable solution from Ford. While some electronics system failure is merely annoying, a miscue of the transfer case could make the vehicle highly unstable by causing the tires to in effect run at very different speeds. If the vehicle were on hard pavement at highway speeds, it would sway violently because of unusual, unnatural torque being placed on it, and its tires would be placed under unimaginable stress.

Ford's decision to assemble every possible function into the GEM control unit means that it must swallow a wide variety of disparate inputs, reject corrupted data and withstand shocks to the system from faulty connectors and faulty sensors. At the same time that the GEM is concerned with making sure that the wheels receive power correctly, it also must decide how fast to run the windshield wipers, whether to keep the dome light on, whether to engage the alarm, etc.

Burned into the GEM's central processor memory are logic sequences designed to prevent any malfunction. For example, a definite logic sequence must occur before shifting into four-wheel drive low is allowed, including a speed sensor signal telling the GEM that the Explorer is going less than three miles per hour. If you believe Ford's line, Explorers must be nearly stopped before the four-wheel drive engages. But this begs two questions—how the GEM circuit knows the vehicle is stopped, and if it ever makes a mistake. The first question is simple—the Explorer is wired with tiny sensors in its transfer case and on the wheels so that it knows how much power is being sent to the wheels and how fast they are turning. The second question is more subtle, but is answered conclusively with a yes.

Please note that the GEM design and computer program information vary not only with every model year, but within years and with every type of Explorer sold. Go to any Ford dealer and tell them you need a GEM, and you will be greeted with a long list of questions. Basically the only way to get the right one is to have the old one in hand. Each year, Ford has continuously added new functions to the GEM while correcting latent problems in the unit. Also, Ford has tweaked the GEM to give the Explorer smooth and effective four-wheel drive performance. For example, the GEM module engages the four-wheel drive during acceleration in both the forward and reverse directions for a short time so that the vehicle does not lose traction. Young drivers will not be burning rubber, but this can cause unpredictable results for everyone.

Explorer Four-Wheel Drive Operation

In older vehicles, power was sent to one wheel, while the others spun freely. Better traction is often needed, and innovations such as limited-slip differentials and four-wheel drive variations were developed in response. But four-wheel drive vehicles place unique stress on tires. The central danger in implementing these innovations is that on hard pavement the wheels will suffer from asymmetric torque where and when they have to spin at different speeds.

Please keep in mind that when any vehicle turns, the front wheels must travel farther than the rear ones, since the actual turning circle for the outside front wheel is larger than for the inside rear (each wheel travels at a different speed through a turn, and the difference depends on the tightness of the turn).

In the Explorer's "four-wheel drive high" mode, the clutch in the transfer case is fully engaged. The front drive train is locked "in step" with the rear drive train, that is, the front axle turns at the same speed as the rear. If you try to turn on hard solid ground, such as wet or dry road surfaces, binding occurs in the driveline as the front wheels try to turn faster than the rear but cannot. Something has to give, and it could be the tires sliding and jumping across the pavement, or it could be a broken axle shaft, drive shaft, or other driveline part. This is why drivers should not use four-wheel drive high on the road unless it is quite slippery.

The same holds true for the "four-wheel drive low" mode. A different gear ratio is used in four-wheel drive low to allow more power for steep hill climbing and descending in low range. As with four-wheel drive high, the front and rear drive trains are locked together through the electric clutch in the transfer case. Consequently, both four wheel drive modes (all-wheel drive) give a permanent fifty-fifty power split from front to rear.

When the system is engaged in the "auto mode," things get more complicated. While the rear axle is the primary drive axle, the front axles do have some power applied as well. This is controlled electronically. The GEM continually pulses the electric clutch such that it is engaged for short bursts, long enough to allow some drive forces to go through the front axle, but short enough to relieve the binding that can occur when turning. These pulses can be fast enough to engage and disengage the clutch several times a second. When traction is good, these short periods of engagement occur such that the front axle is engaged approximately five percent of the time. That is where Ford gets the advertised 95/5 percent rear/front power distribution. When things get a little more slippery, the system reacts to change the power distribution.

The wheel speed sensors that are part of the anti-lock brake system are used to detect rear wheel spin. If the computer senses that the rear wheels are turning faster than the front wheels, the computer automatically tells the clutch to be engaged for a longer portion of the time. This is done in 10 percent increments, up to a duty cycle of 100 percent increments. (It will go to 15 percent front/85 percent rear and so on until traction is regained and the wheel spin stops.) The change begins to occur after within about one-fifth of a turn of the loose or spinning rear wheel, as determined by wheel speed sensor variations.

By increasing the engagement time in small increments, the system can apply just enough torque to the front axle to give added traction, while allowing periods of relief for the drive train to prevent binding. As these engage/disengage pulses occur several times per second, it appears seamless to the occupants of the vehicle. Once traction is regained and the wheels are turning at the same speed (the rears are no longer turning faster than the front wheels), the procedure reverses itself, once again in 10 percent increments. This continues until the power distribution is back to the five percent front/95 percent rear ratio.

There are a variety of failures, discussed below, that would cause the GEM to miscalculate the torque that should be sent to each wheel during highway driving. The driver might not detect this during vehicle operation, but the tires would be placed under high lateral stress that would lead to their eventual disintegration or catastrophic failure. Further, during turns at any speed, an Explorer with a malfunctioning GEM system would also place extreme stress on the tires. Highway-speed maneuvering such as lane changes, and swerving to avoid hazards would also place tires under considerable stress.

We have seen a number of Explorers in the junkyards that appear to have been wrecked by loss of control and subsequent rolling over. Based on what we have observed, the following scenario might occur. Vehicle A (driver A) is driving rather fast and a slower vehicle (Vehicle B, driver B) is merging onto the same lane. Driver A, being the higher speed vehicle, might need to make a quick lane change (by veering to the left to miss the merging vehicle B). The low steering effort of modern power steering systems, and the urgency of the situation may cause driver A to over-steer and veer very sharply. The next step for driver A is to correct or turn to the right. Due to over-steering, this may also require correction and driver A will turn left again.

During this time, a GEM malfunction would cause the wheel speeds to be locked together. As the tire paths are of significantly different lengths at different times, they would attempt to spin at different speeds, but could not. The end result is that high tire stress would result, and traction would be compromised.

Findings

This review determined that Ford's implementation of an electrically shifted transfer case controlled by the GEM system, can fail in a manner that causes Explorers to become highly unstable while driving, and user complaints listed below indicate that this has happened. Further, other electronic failures such as those associated with the Powertrain Control Module (PCM) and the anti-lock brake system (ABS) have caused critical vehicle instability. This has been determined through multiple sources, including exhaustive reviews of Ford technical information, bench tests of relevant components, a simulator controlled operation of a transfer case, and actual road tests of Explorers under induced failure conditions.

GEM malfunctions would occur often, though not always, with a driver unintentionally selecting four-wheel drive by turning the selector switch. Due to the location and similarity of the dash panel controls on some years, this is a highly likely event. A driver keeping his eyes on the road and trying to adjust radio volume or fan speed could easily turn the four-wheel drive selector switch instead.

During the bench analysis of the system, we found multiple hardware failures that could cause unintended shift to four-wheel drive. These were confirmed with actual road tests on the 1998 Explorer Sport. These conditions are:

Simultaneous input failures of the following signals, accompanied by selector switch mis-selection or failure:
- Starter Enable
- Neutral Sense
- Vehicle Speed
- Front and Rear Axle Speed
- Brake On Signal

Shift actuator motor shorted to a power source. This could be caused by:
- Actuator relay failure
- Wiring damage
- Connector malfunction

Short to ground of the relay actuator signal. This could be caused by:
- Actuator relay failure
- Wiring damage
- GEM Failure

We found numerous software/processor errors that could result in unintended range change:
- Power transient (glitch) causes GEM module to incorrectly read input signals. This could cause an incorrect decision to change range.
- RFI (Radio Frequency Interference) causes GEM module to incorrectly read input signals. This could cause an incorrect decision to change range.
- Program counter corruption (caused by power transient or RFI) branches to Range Change Routine. This could force a range change.

There are several observations related to this study:

- Most or all of the inputs to the GEM are analog (variable levels) depending on the unit. An example is the four-wheel drive Selector Switch, which supplies a different voltage level to request different operating modes. At some point signal degradation from a host of sources, or low voltage conditions could affect GEM operation.

- The GEM module is enclosed in a cheap plastic housing which does not appear to have any RFI shielding, and provides no protection against moisture. In addition, the circuit board is not conformal coated. Conformal coating provides a moisture barrier and inhibits the degradation of the circuitry due to corrosion, fungus growth, or component chemical discharge. Standard military type environmental testing for electronic assemblies, such as defined in the test methods of the military standard MIL-STD-883 (Test Method Standard for Microcircuits), could be used to accelerate the propagation of such failures. The usual mounting techniques employed on the GEM offer no vibration isolation or mechanical shock protection. Excessive vibration or shock stress can damage components and crack solder joints resulting in intermittent, erratic operation, or complete failure of the circuits.

- Computer processors inherently perform operations serially (one at a time). The GEM is no exception and performs multiple functions (multi-tasking) in a serial fashion. By doing them fast enough (in sequence) the module appears to be performing the operations all at once. We did not attempt to overload the GEM processor; however, all systems are subject to exceeding the available amount of resources.

- While manipulating the Start Enable and Neutral input signals to the GEM for testing purposes, the GEM responded as if to a tone request (d test or diagnostic function), indicating that it can be confused and respond incorrectly.

- Because the inputs are analog, the GEM should have been designed to sense open (broken) wires. This was not done as evidenced by a lack of response to open sensor signal conditions.

- The GEM wheel-spin-control algorithms are not always correct for given situations. There are thousands of possibilities with the variables involved. An example of this was reproducing a skid on wet pavement resulting in increased power to the front wheels, and consequently more wheel-spin. The opposite of the desired result, this actually reduced steering control, which is necessary for skid recovery. This can be repeatedly demonstrated by skidding the vehicle sideways on a wet (low traction) street and applying power.

Our reviews of Explorer user complaints and interviews with Explorer owners who have suffered from vehicle failures have documented repeated failures in the manner described above, from a design point of view. Information obtained from accident and near-accident victims also indicates that Ford has failed to enter these problems into its repair database, despite clear evidence and acknowledgements by mechanics of their existence. For instance, one Explorer driver posted this statement on an Internet bulletin board:

> I have a 2002 XLT that suddenly put itself into 4 wheel low range while my wife was driving at approx. 40 mph. Fortunately she was wearing her seat belt. She said it stopped so suddenly she almost hit the steering wheel. The dealer claims the computer sent a faulty signal, (no kidding), to the transfer case.

We found other clues in owner postings on the Internet. Owners have complained about excessive tire wear, which would be a pre-accident indication of GEM malfunction. For example, one woman recently posted this complaint:

> I just purchased a '98 4x4 Eddie Bauer. The vehicle looks well cared in all aspects except the tires. They have a lot of tread left in general, but severe wear on the inner and outer tread. I would call it cupping, but most cupping I have ever seen is only on the outer tread. My tire guy said it was probably mis-alignment that caused it. I'm just concerned that there is an inherent suspension related problem with the Explorer that is difficult to fix. I feel I checked the vehicle out in every aspect but this one. Does this problem sound familiar to anyone?

We have found also separate owner complaints related to all other elements of the GEM system. For example, this complaint concerns the critical Vehicle Speed Sensor (VSS) component. If the VSS fails, the Explorer thinks it is stopped and will enter the four-wheel drive mode when other conditions are met. The owner wrote:

> Some time ago I noticed the speedo(meter) dial is shaking in a strange manner. It gets worse and worse until there are problems in 4WD Auto—the computer couldn't manage with speed difference inputs from front and rear axles and switched 4WD off and (the diagnostic trouble code) was "VSS intermittent." A test with (a) VSS taken from another Explorer proofed that it's a VSS (problem) indeed and so I ordered a new one.
> In the very beginning it seemed to be OK but very soon (a day or two) the dial started to tremble again. Today I did a bit longer trip (partly off road) and as result I've got the VSS in the same condition as it used to be! I have no ideas what is the reason. Has anyone experienced the same thing or perhaps can offer a convenient explanation what the hell is going on. Explorer is 95 4L OHV. Any advice is appreciated.

Though the GEM disabled the four-wheel drive in this case, we have documented conditions where the four-wheel drive would remain engaged and place the Explorer's tires under stress.

Methodology

My work in researching this book focused first on obtaining shop repair manuals, schematics, troubleshooting guides, and diagnostic equipment associated with the Explorer series of vehicles. Examining these sources allowed me to catalog a series of questionable design decisions Ford engineers made when pushing the Explorer into production. My assistants and I then examined production vehicles and built a test system from components obtained from wrecked Explorers.

We found a series of circuit design decisions that, while they appear functional on a schematic, were not examples of competent vehicle system engineering. In the realm of vehicular design, merely "functional" is not sufficient given the requirements at hand. Explorer electronics design is characterized by an "add-on mentality," where circuitry for new features is placed on top of old systems. Wiring, connectors, and even the GEM itself were placed in areas where they would be subject to unnecessary heat and moisture. All of these components showed Ford's emphasis on low-cost manufacturing proved to have a high propensity for failure.

On an Explorer, replicating a GEM malfunction during vehicular operation is highly dangerous, and this risk was mitigated in several ways. We built a bench test system to allow us to document component operation (which was sometimes different than in Ford documentation) and simulate failures. Our vehicle road tests were conducted at speeds of 35 mph (56 km/h) and lower.

An important part of our investigation concerns integrating the results of our tests with accidents and near-accidents experienced by Explorer drivers.

My assistants and I developed the following information sources to create our model describing GEM Explorer failure:
- Schematics/Designs
- Internal Ford Documents
- Bulletin Board Explorer Owner Complaints and Inquiries
- Device Bench Testing
- Road Tests with Modified and Unmodified Vehicles
- Interviews with Explorer Owners
- Email exchanges with Ford Mechanics
- Consultation with non-Ford mechanics
- These sources provide an effective information base to assess and describe Explorer failures. Each element of this research supports the oth-

ers. For example, through our bench testing, we discovered errors in Ford documentation with respect to electronics implementation. This deviation from design made the Explorer more prone to failure, and the incorrect documentation make it likely that mechanics have missed vehicle problems.

For all possible events that we discovered through investigating the documentation, components, and vehicles—we found the actual event occurrence described in owner complaints.

Technical Data Review

The scope of work centered on learning how the GEM module works, which systems/components interface with the GEM, and what Explorer features are relevant to GEM-related failures.

Our analysis was hampered initially by Ford's generally poor documentation relating to the GEM and how to troubleshoot its problems. The lack of a single GEM design, even for a particular model year and vehicle type, complicated the effort. We were able to obtain a representative group of GEMs, related sensors, schematics, and manuals. These were utilized to establish the device's failure mechanisms.

Though our work is centered on the GEM's four-wheel drive function, all functions are relevant to understanding any system failure. The key point concerning GEM failure is that Ford's decision to combine critical with non-critical functions in a single unit compromised the reliability of the device. The non-critical functions create opportunities for the GEM to fail because they have connections that could become pathways for electromagnetic interference. These functions also require processor multiplexing that is an important opportunity for failure.

The anti-lock brakes (ABS) used on the Explorer are controlled by the GEM and also present failure opportunities. If the ABS sensors individually or jointly lose communication with the GEM, the default option is for these brakes to engage fully when the brakes are applied. This can cause one or more of the wheels to lock, placing severe stress on the tires. The Explorer's automatic load leveling function is also controlled by the GEM and has been evaluated here. The electrical transfer case used in the Explorer received considerable analysis. The Bronco began using the Borg Warner 1350 Electrical Transfer Case in 1986, and the Borg Warner 1354 was included in the Explorer when the line began in 1991.

The GEM has proven to be an irresistible place for Ford to add new Explorer features, as the GEM approach has reduced the cost of implementation. The feature count has grown such that, in addition to the complex four-wheel drive controls, the GEM on the 1999 Explorer has these connections:

- Front windshield wiper/washer mode input
- Liftgate windshield wiper/washer mode input
- Interval/delay wiper input
- Wiper speed control
- Washer pump control
- Battery saver relay control
- Defrost switch input
- Interior lamp relay control
- Brake sense input
- Door ajar indicator control for all two/four doors and liftgate
- Safety belt warning
- Throttle position switch
- Message center for Diagnostic Trouble Codes
- Main light switch
- Powertrain Control Module Interface
- Instrument Cluster Interface
- Power Windows
- Air Suspension Control Module Interface
- Remote Anti-Theft Personality Module
- Door Disarm Switches
- Digital Transmission Range Sensor
- Electric Shift Control
- One-Touch Down Window Relay Control
- Accessory Delay Relay Control
- Park sense
- Key in ignition input

Just as running too many software applications on a personal computer will cause them to function more slowly or to freeze occasionally, Ford's addition of these features creates a situation where the GEM resources might be overloaded. As a sampling of user complaints presented below demonstrates, devices connected to the GEM demonstrate unusual failures.

Event Modeling

Based on our study of the Explorer and available documentation, we created several theoretical events that could replicate catastrophic GEM-related failures. One failure event selected as valid was an uncommanded shift from two-wheel drive into four-wheel drive high or low at city driving speeds (35 mph or 56 km/h). The events related to this failure are discussed above in the Findings section.

This scenario is based upon the Explorer's four-wheel drive system operation. When in automatic four-wheel drive, the generic electronic module (GEM) controls power to the transfer case electric clutch in response to wheel slippage, as seen through the front- and rear-drive shaft speed sensors.

On Ford's Control Trac four-wheel drive system, the driver can select three different modes of operation, including automatic four-wheel drive. The electronically shifting transfer case consists of front and rear magnetic field-style drive shaft speed sensors, electric four-wheel drive shift motor, and an electric clutch.

The electric clutch is used to engage the front drive shaft to allow shifts into four-wheel drive at any speed—even when the front wheels are stopped, and when the vehicle has become stuck while still in two-wheel drive. When in low range mode, the electric clutch is fully engaged. In 1995, the Explorer went to the GEM module to help with the electronic four-wheel drive systems. Ford's ESOF (Electronic Shift on the Fly) system relies heavily on the GEM for proper operation.

The Electronic Shift on the Fly (ESOF) system is an electronic shift four-wheel drive system that allows the driver to choose between two different four-wheel drive modes, as well as two-wheel drive. The driver can switch between two-wheel drive and four-wheel drive high mode at speeds up to 55 mph (89 km/h). To engage or disengage four-wheel drive low range, the vehicle speed has to be below three mph (five km/h), the transmission in neutral, and the brake pedal applied.

The transfer case is equipped with an electromagnetic clutch inside the case. This clutch is used to spin up the front driveline when shifting from two-wheel drive to four-wheel drive high mode at road speeds. When you turn the control switch on the instrument panel to the four-wheel drive high mode, the GEM recognizes that a mode shift has been requested. It then activates the electromagnetic clutch and the transfer case shift motor relays.

When the shift motor reaches the desired position (as determined by the contact plate position signal), the GEM removes power from the shift relays and motors. When the transfer case's front and rear output shafts are synchronized, the spring-loaded lock-up collar mechanically engages the main shaft hub to the drive sprocket. Finally, the front axle collar engages, and the GEM deactivates the electromagnetic clutch, locking the front and rear drivelines together.

Explorer Systems Involved

Study of the GEM or any vehicle computer in isolation is useless for understanding the electronics system. The key to determining the extent to which a GEM malfunction would cause overall vehicle problems is understanding how the device is integrated into the vehicle and how other design flaws contribute to problems. Only when combined with these other flaws—poor sensors, bad wiring, a high center of gravity and improper suspension—do vehicle processor malfunctions lead to dangerous vehicle conditions. This report deals with those other factors briefly but in sufficient detail to understand the significance of GEM malfunctions.

The Automatic Ride Control (ARC) system lifts the vehicle when the transfer case is shifted into automatic four-wheel drive (by rotating the transfer case switch) or four-wheel drive low (rotate the selector switch to the four-wheel drive low position, place the transmission in neutral, and tap the brake pedal to engage or disengage). Improper, GEM-directed ARC operation will raise the Explorer's center of gravity, increasing the chance of rollover.

The Powertrain Control Module (PCM) controls transmission operation and is set to limit the vehicle's top speed. Automakers are supposed to limit vehicle top speed with respect to certain safety factors, including the speed rating of tires used as original equipment. Ford recalled Explorers to reprogram the PCM to lower their top speed after experiencing tire failures on OEM types due to overheating.

The anti-lock brake system (ABS) provides critical information to the GEM so it can determine torque applied to individual wheels, as well as for directing four-wheel drive system engagement. Inaccurate, corrupt or missing ABS information will cause GEM miscalculation.

Why Understanding CMOS Latch-Up is Important

Ford recalled 1999 and 2000 Explorers and Mountaineers for a GEM malfunction because the system could experience failure referred to in this case as "lock-up." More properly called "latch-up," this phenomenon affects microcircuits made using complementary-symmetry metal oxide semiconductor (CMOS) technology. A surge of voltage and current on a computer input or output causes CMOS circuits to stop functioning, which in the case of the GEM, could cause total failure. But, once power is removed and re-applied on a CMOS device in latch-up, it reverts to normal operation without any evidence that it failed. Since GEMs are failing because of latch-up rather than burn out, other investigators have missed this cause of failure.[5]

The symptoms of GEM lock-up were that some GEM-controlled functions— front windshield wipers, interior lights, and the four-wheel drive system—could not be turned on, or in some cases, turned off which indicates a GEM output that is unrequested and cannot be changed by its normal control unit. While it should have been apparent to Ford and the National Highway Traffic Safety Adminis-

---

[5] As we discuss in this book, there are six major failure mechanisms in Integrated Circuits: 1) localized high temperatures; 2) localized dielectric breakdown; 3) metal corrosion; 4) electro-migration of the metalization; 5) surface inversion caused by ionized contamination in the passivating silicon oxide layers, and 6) CMOS latch-up. Of these, CMOS latch-up is the most difficult to detect after failure.

tration (NHTSA) that the GEM could have been sending the four-wheel drive system erroneous signals during highway driving, no association was made with the GEM and the Explorer's catastrophic problems. NHTSA's assumption/contention that the four-wheel drive system was merely disabled was inaccurate.

The symptoms of GEM lock-up were that the GEM computer would issue improper codes, which caused unrequested system operation during the lock-up condition. This was experienced by many owners when the windshield wipers turned on and could not be turned off, interior lights turned on and could not be turned off, and the four-wheel drive malfunctioned in various ways including a shift into four-wheel drive low unexpectedly. The nature of a lock-up on a CMOS chip is that it causes the chip to stop functioning, but allows the chip to assume a static condition where the output signals (codes) can assume a random static condition until power is removed. The chip can then resume normal operation.

The fix ordered by NHTSA on the GEM recall involved only installing a two-cent resistor on the device.[6] Putting a resistor on the GEM will let the excess current flow out and end the latch-up situation, but not prevent it. By the time normal GEM operation is restored it will have been malfunctioning for a significant amount of time, more than enough to have miscommanded the four-wheel drive function or some other function.

Forensic Examination

Key to our work was the examination of numerous Explorer and other menu features components obtained from salvage yards. Unlike wrecked vehicles that are impounded by the government or kept by plaintiffs' lawyers, disassembly on vehicle salvage yards begins within a day or two. To get a good look at such vehicles and obtain desired parts, we established working relationships with vehicle dismantlers that allowed us immediate access to wrecked vehicles at several salvage yards.

We examined and obtained numerous parts for our simulator from a 1998 Explorer that was wrecked in a single-vehicle rollover accident. From this vehicle, we took the transfer case, GEM, sensors, and cabling. We also obtained GEMs and other components off the shelf, and from other Explorers in salvage yards. This allowed us to study variations in GEMs and in their installation. The GEMs we examined were ones from various Explorer model years, as well as one from a Ford Ranger for comparison.

Among the GEM modules analyzed were:
- 1995 2WD Module F67B-14B205-AA (Ranger)
- 1996 2WD Module F77B-14B205-AA (Explorer)

---

[6] NHTSA Recall No. 00V072/Ford Recall No. 00S04

- 1998 four-wheel drive Module P1F87B-14B205-CC (Explorer)
- 1998 four-wheel drive Module UN150 (Explorer)

These GEM variants provide the basis for the following discussion. Please note that the functions of different modules were incorporated into one device, the GEM, in 1995, and Ford added new functions with each model year.

Ford seeks to differentiate its SUVs from the competition with electronic features that provide perceived high value for the consumer. For example, with the 2003 model year Expeditions, Ford offers seats that fold with the touch of a button. Like the electrical transfer case, this appeals to soccer mom drivers for ease of operation. It also appeals to a whole generation of people who grew up with pushbutton technology in consumer goods, from remote controls to microwave ovens.

The constant addition of new features, and an apparent wide variety of initial designs, combined with field alternations to the GEM, creates a challenging technical environment for Ford owners and mechanics because the GEMs can fail in so many different ways. Due to the realities of service economics, and the lack of documentation, the favorite repair technique for many Ford mechanics when there is a GEM related malfunction, is to simply install a new GEM. Since many GEM problems occur when other elements of the system are failing, such as sensors and wiring, this technique often fails to eliminate the problem,

There are few good locations on the Explorer, and many bad ones, for electronic components such as the GEM. Ford chose one of the poor locations for the Explorers, as well as the Ranger trucks—directly behind the radio in the dash board. This not only subjects the GEM to intense heat from the sun through the windshield, from the heating system, and from the radio, but also to water that leaks in along the radio antenna line and other sources.

The GEM's basic fragility in the dashboard location is illustrated by the 1995 GEM. This unit, which uses a National Semiconductor processor, operates only the windows, wipers, and alarm system. This specific example was removed from a 1995 Ranger that had experienced GEM overheating that resulted in uncommanded windshield wiper operation, and power windows being inoperative.

The 1996 GEM from a two-wheel drive Explorer allowed for further study of the GEM implementation as isolated from the four-wheel drive function. This unit has a Texas Instruments processor.

Please note also two 1998 GEM units: one is from the wrecked 1998 Explorer shown above, while the second is from a salvage yard. Both use Texas Instruments processors, but one uses an EEROM programmable microcomputer. This is a significant design difference considering they are from the same year and model. The conclusion is that Ford was unwilling or unable to freeze the GEM's design because of performance problems with the device and/or a desire to cut its cost on an ongoing basis while including more features.

Bench Test Simulation

Our work in understanding the GEM and its malfunction modes was advanced by our construction of a bench test simulator, which proved to be a time-consuming process because of the errors in Ford's documentation. The discovery of these errors is important for understanding why GEM problems were not detected or corrected.

The initial activity on the project was to construct a four-wheel drive transfer case simulator. During early tests the simulator proved unable to change range in the transfer case due to some unknown missing signal not included in Ford documentation. This necessitated acquiring a vehicle for further study. The missing signal was found to be the starter enable signal (which required a starter to be present). The simulator was modified to actuate that signal, resulting in action of the transfer case that correctly mimicked the actual vehicle operation.

This GEM starter interlock feature works as follows. The starter relay requires ground to one end of the relay coil and application of +12 V to the other to start the engine. Ground is supplied to the relay through the Neutral Safety Switch in the automatic transmission. This switch only supplies ground to this relay when the automatic transmission is in Park or Neutral. This protects against someone starting the engine when the automatic transmission is in drive or reverse. The ignition switch provides +12 V to the positive end of the relay when starting the engine (in the "start" position). A sense wire is connected from the +12 V side of the relay to the GEM module.

Fig. 35. **The Borg Warner transfer case obtained from the wrecked 1998 Explorer equipped with Michelin tires.**

The GEM measures +12 V on this wire when the engine is being started, but measures ground when the transmission is in Park or Neutral. Ford also has an extra switch in the automatic transmission that is closed when the transmission is in neutral.

Neither ground nor +12 V is connected to the wire for any other operation. What this means is that electricity cannot flow to either the negative or positive power source—it is "dangling" and can only send electrical shocks back to the GEM. This design flaw creates an antenna on the GEM that attracts electromagnetic energy to the module. This energy could latch-up the GEM processor.

The presence of ground or +12 V provides another check to the GEM that the vehicle is not moving. This wire is not shown in the documentation from Ford. This method of informing the GEM that the transmission is in park or neutral is operationally weak. A short in the wire would provide a false park/neutral signal to the GEM.

The Simulator/Transfer case was tested to try to initiate a range change under moving vehicle conditions. Although the transfer case attempted to change range, it would not change ranges until drive power was removed. We speculate this is a result of the driveline torque (gear lash) preventing the change until the gear lash approached zero. In vehicle operation, gear lash disappears momentarily when beginning acceleration and deceleration, or in between during coasting. This would allow for the transfer case range shift, as demonstrated in the vehicle tests described below.

Why We Conducted These Tests

The bench tests provided the necessary bridge between the documentation study and the live vehicle tests. They allowed us to isolate individual Explorer electronic functions and examine each component carefully in operation. Ford's schematic for GEM implementation could be compared side-by-side with the GEM system to detect deviations. These tests also allowed us to conduct our research without the danger of inducing an actual Explorer rollover during the tests by providing an understanding of what to expect during the live vehicle tests.

Equipment Used

The GEM system was removed from the 1998 Explorer shown and reassembled in the bench test environment. We used several pieces of diagnostic equipment, including the Diagnostic Trouble Code (DTC) reader used by dealerships and an oscilloscope to analyze specific signals.

Fig. 36. **Front view of wrecked Explorer**

Fig. 37. **Side view of wrecked Explorer**

Setting up the simulator required considerable engineering and technical resources. From wrecked Explorers, we got the transfer case, the GEM unit, the wiring and connectors to interface to the GEM simulator system, and the test equipment. We also required a variable speed motor and a shaft that would fit into the transfer case to simulate driving conditions. We found the correct spline shaft at a local transmission shop.

For the motor function simulation, a variable speed electric drive motor was used. It has a maximum no-load speed of about 3,000 RPM, and was controllable to provide a range of RPM to simulate the engine and transmission under various conditions. The motor was connected via a flexible coupling to the shaft. The first test of this configuration was a temporary hookup of the mechanics (to prove they worked). The drive system indeed worked. The next step was to mount the parts on a secure base and connect the simulator system.

The specific identification and patent numbers on the transfer case are:
Borg Warner Transfer Case
Metal ID tag:
J1972 4405-10
129754 F87A-CA
ID label on Casting:
GMAZ91D
8CA
44-05-000-010
BORG WARNER AUTOMOTIVE
F87A-7A 195-CA

Tests Conducted

We conducted extensive tests to define and validate the universe of GEM system failures and consequential events. For each component, we simulated each possible way it could fail and measured GEM effects. This was done both individually, and in groups and stages. The results of these tests are described below.

What We Found

Our tests revealed the following problems with the four-wheel drive Explorer:
Transfer Case sensor failure: This failure causes the transfer case to NOT change from low range to high range even though the mode selector switch and light indicates the change. This failure can occur on the Borg Warner 44-05 or C-Trac transfer case used with the electronic mode selector. The 44-05 with electronic mode selection uses a DC motor to change from high range to low range and back. The motor direction is controlled by a dual coil relay that is operated by the GEM module. The GEM module received input from the mode selector switch for the desired high or low four-wheel drive range, or a four-wheel drive in 1997 and later models. Certain conditions must be prevalent before the GEM module will command the four-wheel drive condition. These conditions include correct inputs from several sensors before the four-wheel drive range selection is executed. When these conditions are met, half the dual coil is operated and the motor turns in a specific direction. The

transfer case uses a four-phase feedback system to the GEM module to tell when the mode selection is completed. Four wires and ground provide the output from the transfer case in the form of pulses indicating where the transfer case range selector motor is positioned. If this system gets out of adjustment, the transfer case can be in the wrong range, while the GEM module input will appear correct. This is a fault condition.

Inadvertent four-wheel drive: This can be caused by failure (shorting) of the solid-state relay, or a speed sensor missing pulses. When the speed sensor malfunctions, the GEM module can decide that the front wheel(s) are slipping when they are not. The Borg Warner 44-05 or C-Trac transfer case uses an electromechanical clutch to shift the front driveshaft. The GEM module operates an electronic (solid state) relay to activate the clutch. If the electronic relay should short circuit, the front wheel drive will be engaged regardless of other conditions. This means this can occur independently of the mode switch selection. Sensor problems can also provide incorrect information to the GEM module, resulting in malfunction.

1998 Explorer windshield wiper/washer interference: The 1997-1998 Explorers change the wiper speed based on vehicle speed (which is determined by input from the VSS). Circuit malfunctions can cause the transfer case to NOT go into four-wheel drive low range. This can occur because the return lead is common between the GEM module transfer case speed sensors, the wiper/washers, and the four-wheel drive mode switch. Malfunctions in these interconnected circuits can cause problems in operation of the transfer case.

A four-wheel drive during acceleration: The GEM module engages the front driveline clutch during acceleration in both the forward and reverse directions for a short time. This can cause unpredictable results.

Lack of four-wheel drive: It appears that the GEM module disengages the electric clutch before changing into low range. If the change to low range does not occur, the vehicle may be stuck in two-wheel drive low-range (a fault condition) without the driver knowing.

Understanding Hidden Design Defects

The weak engineering used in the GEM system's design contains fail points that would not be obvious to some investigators. Further, a mere examination of schematics by inexperienced investigators might determine that sufficient fail-safe features exist to prevent improper range shift. The use of an actual GEM system outside the Explorer provided necessary visibility into the system's faults for demonstration and documentation.

Vehicle Test

Based on technical information developed in previous program work, we moved from bench tests to testing with modified and unmodified vehicles. This work allowed us to create visible proof of the problems

Validity of Test Vehicles

Our primary test vehicle is a 1998 XLT Sport Explorer. We modified this Explorer as described below to test key elements of our GEM theories. The 1998 XLT was selected because that model year had a large number of electronic features. Our XLT was well equipped. Having the XLT version allowed us to evaluate a vehicle with all luxury features interfacing with the GEM. These include the power windows and locks not included on the XL.

Test Events and Results

The vehicle was modified to manually control the signals that prevent range change while the vehicle is in motion. These signals consisted of Starter Enable, Neutral Sense, Vehicle Speed, and Front and Rear Axle Speed. The start input was grounded, the neutral input was grounded, the vehicle speed sensor was disconnected at the GEM module, and transfer case sensor power was disconnected to disable the transfer case speed sensors. The last required signal for transfer case shift is the "brake on" signal. This was left connected to be used to manually trigger the shift.

Low-Range Shift Test

Test Conditions: The vehicle was driven at a constant 25 mph (40 km/h) and the transfer case was manually switched from auto (high) to low range. The switch was made under power, and then the brake pedal was slightly depressed (to activate the brake on signal). The vehicle was driven at approximately 25 mph (40 km/h) with the brake depressed (no braking action) and power applied until the transfer case engaged four-wheel drive low range. The test was performed on an isolated city street, which was relatively new, and in excellent condition. There were no surface anomalies to interfere with the test, i.e. the surface was smooth.

Test Report: Two runs were made for the first transfer case test.

First Run: The first run confirmed that the transfer case would engage at speed with the test conditions as described. The vehicle was run at 25 mph (40

km/h), the four-wheel drive switch was moved to the four-wheel drive low range while maintaining throttle position, and the brake was actuated to initiate the shift. The transfer case engaged the low range at approximately 23 mph (37 km/h). This occurred at about four seconds after switch actuation.

The switch was hard, with the resultant gear ratio change affecting both speed and RPM. After the switch, the vehicle was slowly braked to a stop. The switch was returned to the auto position, the brake was released and re-applied to effect the switch to auto four-wheel drive mode. The vehicle was turned around (180 degrees to set up for the second run). During the turn-around, we encountered some unknown mechanical interaction from the driveline. The actual action is unknown, however it manifested itself in a "clunk" from the drive line, similar to low range engagement. It is speculated that the transfer case had not fully returned to the high range.

Second run: The second run proceeded as the first, with the purpose of checking RPM variance when the low range engaged. The test proceeded as before, and the engagement was similar. On this run the shift occurred as before, at approximately four seconds after the switch was changed and the brake applied. RPM was observed to increase from 1,600 rpm to about 3,000 rpm.

The effect on the engine speed is somewhat gradual, probably due to slippage in the automatic transmission. The damping of the tachometer may also effect the speed at which the indicated RPM is shown. It must be noted that at the time of the shift, with the resultant gearing ratio change, it is difficult for the driver to maintain a precise vehicle speed. The gear change does affect vehicle speed, requiring changes in throttle position and braking force in an effort to maintain the prescribed test speed of 25 mph (40 km/h).

Observations: After running this test several times, it was observed that shifting to low range (with the logic modified as referenced) is definitely possible. The vehicle will shift at speed; however, the actual shift seems to be somewhat dependent on driveline loading. Torque on the driveline seems to inhibit the shift. This is probably similar to gear engagement in conventional standard shift transmission without the benefit of synchromesh. Once the factors of torque, and gear (vehicle) speed are within certain parameters, the shift will occur. If these could occur within certain parameters, it would appear that the shift could occur at higher speeds.

Our third test involved the anti-lock brake system (ABS) used as a GEM input. The right front wheel speed sensor was disabled. At start-up, the computer displayed an ABS warning light. The diagnostic system was connected. The vehicle was driven on city streets at speeds up to 40 mph (64 km/h). The vehicle drove normally, except that it was possible to lock at least one wheel during hard braking. The computer displayed no other warning lights, and set no codes.

The left front wheel speed sensor was then disconnected. Again, at start-up, the ABS light was on. The vehicle was driven as before, and drove normally. As before, it was possible to lock at least one wheel during hard braking. The computer displayed no other warning lights, and set no codes.

At this point, the test was concluded.

We also conducted other tests related to the Explorer's anti-lock brake system (ABS). Initially the Explorer was driven with diagnostic equipment connected to verify sensor operation. Both front wheel speed sensors were functioning normally. Each speed sensor was then defeated individually, left then right. This was accomplished by opening the circuit at the connector, simulating a failed sensor, broken wire, or bad connection.

After the initial startup sequence was completed (left sensor defeated) the ABS warning indicator stayed on, as did the airbag warning light. The ignition was cycled off and on and the engine re-started. This time the air bag warning went off as usual, and the ABS warning light stayed on. The vehicle was driven a short distance to confirm the left front speed sensor read zero. This was confirmed.

The process was repeated with both sensors disabled. As before, the ABS warning light stayed on, although the airbag light went off after its prescribed time on the first start. The sensors were confirmed to read zero. There was no change from the previous test. No driving anomalies were encountered.

We also conducted tests of the anti-lock brake system (ABS) to determine if likely failures would stress tires as well.

The vehicle was driven on a road similar to any city street in good condition. The surface was concrete. The vehicle was driven at 35 mph (56 km/h) and stopped with maximum braking force with all sensors operating to test the anti-lock brake system. The system is very noticeable when it is active. During the braking phase, the individual wheels will hit lock-up (producing a chirp from the tires) and then begin rolling again as the ABS controls the braking. These short periods of lock-up last only a few milli-seconds, and occur randomly around the four wheels. You can actually feel the system working.

After the initial test, the right front speed sensor was disabled. As before, the ABS warning light was displayed. The vehicle was driven at 35 mph (56 km/h) and then maximum braking was applied. On the first run, the two rear wheels locked. The right front wheel locked, and the left front did not. The vehicle skidded with the rear end swinging to the right. The amount of pedal force was the same, but the pedal travel was increased. On the second run, the speed was increased to 45 mph (72 km/h). At maximum braking, all four wheels locked solid until the vehicle came to a stop. The vehicle skidded in the same manner as before. This test was repeated several times with the same results. No ABS action was observed or felt.

There was a second test of the engagement of four-wheel drive low range with GEM Module sensor inputs disabled. For this test, the neutral input was grounded, the start input was grounded, the vehicle speed sensor was disconnected at the GEM module, and transfer case sensor power was disconnected to disable the transfer case speed sensors. The last required signal for transfer case shift is the brake on signal. This was left connected as before to be used to trigger the shift.

The vehicle was driven at a constant 25 mph (40 km/h) and the transfer case was manually switched from auto (high) to low range. The switch was made under power, and the brake was applied. The vehicle was driven at 25 mph (40 km/h) for five seconds (with the brake and power applied) and then the throttle was closed. The vehicle was allowed to slowly brake to a stop.

On the first run, the vehicle did not immediately shift, but stayed in high range until the vehicle speed was around five mph (eight km/h). It then shifted into low range, grinding the gears on the way in. The controls were reset and the test repeated. On the second run, the setup was the same, but less brake and throttle were used to hold the vehicle speed at 25 mph (40 km/h) for the five-second phase of the test. This time the shift occurred while the vehicle was at 25 mph (40 km/h) during this five-second phase of the test. The shift was very hard and very loud, with no noticeable grinding, and slowed the vehicle considerable. To an unsuspecting driver, such an unannounced shift would be quiet dramatic. After the test, the selector was returned to auto, and the vehicle drove normally.

Test Conclusion: After the test, the vehicle was restored to the standard configuration. There was some initial difficulty with the ignition switch sensor reading being incorrect. The GEM module was disconnected and re-connected, and this cleared the fault (suspected latch-up). After this, the vehicle performed normally. This was confirmed with the diagnostic equipment. The vehicle was then driven in a usual "urban" manner for approximately 25 miles; it appeared to function normally.

The four-wheel drive Explorer is a complicated vehicle that uses technology to buffer the driver from road conditions. Drivers who do not take the time to educate themselves on the various systems that affect handling may not make intelligent choices in regards to their driving style and/or how they handle poor driving conditions. When encountering dangerous conditions, the various systems may isolate the driver to the point that when the computer can no longer maintain stability, it is difficult and sometimes impossible for the driver to regain control.

User Problem Report Integration

Correlating our work with actual Explorer owner complaints is important for validating our findings. We l ave done this by monitoring relevant Internet bulletin boards and contacting persons with Explorer problems that are similar or identical to those suspected as being rollover causes.

In one U.S. case, the driver experienced chronic problems the dealer was unable to correct. He wrote:

My 97 Explorer XLT switches into 4-wheel drive without my turning the switch. Usually, but not always, the indicator light will flash before this happens. My mechanic is having trouble diagnosing the problem because it doesn't happen all the time. Is anyone aware of this being a "known problem," and if so what the causes may be?

We contacted Jim and first offered to help him diagnose his problem. He had already spent about $1,000 at the dealer, but the mechanics there were unable to fix the defect. An important data point is that the mechanics said there were no records of complaints for this problem in Ford's database.

Another Explorer owner diagnosed his 1996 vehicle's problem as improper locking of the differential, something that would place great stress on the tires and could be caused by GEM malfunctions. He wrote:

(I have a) 1996 Explorer XLT with Control-Trac. If I go in to four-wheel drive Auto, when I turn I get grinding on front end. Sounds like the differential has been locked by the actuator. When in four-wheel drive Auto, the front end SHOULD have differential action, i.e., no locking. The four-wheel drive Auto should also only engage when there's rear wheel slippage, but should never lock-up (four-wheel drive Low locks up, which is normal).

After I move the switch, I get the four-wheel drive and four-wheel drive Low dash light blinking 6 times every 2 minutes, regardless of switch position. If I shut the truck off and start it again, there's no flashing. Apparently it's triggered by putting it in four-wheel drive Auto or four-wheel drive Low (and it's detecting a problem of some sort).

The four-wheel drive Low seems to work fine (but I'll get the blinking too). The four-wheel drive Auto seems to be operational also, but as I said above there's the grinding when you try and turn. I am sure it's the front differential getting incorrectly locked in four-wheel drive Auto as it normally is when you go into four-wheel drive Low. When I go back to 2WD, I hear the "pop" as the driveline windup releases on the front axle. Per the manual you should be able to drive around on dry pavement in four-wheel drive Auto if you want, with Control-Trac (that's not the case with four-wheel drive Low—you must be on slippery ground like sand, snow, ice, gravel, or you will get the grinding/popping as the driveline tenses up).

I've heard there's an actuator motor or some such on either the front differential or the transfer case that is balky and needs replacement. Is this a DIY, or does a mechanic/Ford have to do it?

There are other user complaints in which the users do not realize that the problems they encounter are associated with GEM malfunction. Please recall that the door locks are a GEM function. This Explorer owner wrote:

I have a 2002 XLT that I park outside. When it freezes after a rain, the driver's door, and sometimes the front right door locks seem to get frozen. The remote, the keypad, and the electronic controls on the armrest do not work the frozen ones. The others operate normally. If I go in through the back door and manually work the lock then I can get in the front. Even after manually working the lock mechanism back and forth a few times, the electronic control still will not lock/unlock it. The electronic controls do start working again after the temperature gets above freezing. It behaves like water is getting in somewhere and freezing in a way that disables the electronic locks.

If the locks themselves were frozen, the owner could not actuate them manually. This malfunction indicates Ford has not solved water-based shorts or low temperature problems affecting the GEM. Another 2002 Explorer owner posted a complaint stating that his vehicle is being affected, apparently, by electromagnetic interference. The symptoms indicate a GEM malfunction.

I have a 2002 XLT 4.6L with probably one of the first Advance-Trac (A-T) systems that was installed. I've had the vehicle since the end of July and now have about 9,200 miles on it. I have my doubts about whether the system is working right, and would like to know if anybody else has had experience with the A-T system.

The owner's manual is not very clear about how this system functions, or what someone should be concerned about. Over the past four months, I have experienced the same "event" occur with the Advance Trac several times, the most recent being this afternoon. After the vehicle has sat for several hours at work, I've started home and within about a minute of leaving the parking lot, there has been a "rumbling" sound, the dash message center beeps, a message appears that the A-T has turned off, the dash A-T warning light goes on and the last message that appears (and stays on) says to check the A-T system.

The system seems to reset itself only when the vehicle is turned off and restarted. The first two times this occurred happened to be rainy days; however, it has subsequently happened under other weather conditions. A rather bizarre coincidence is that each time, this malfunction has happened within a hundred feet or so of the same spot. The owner's manual suggests that the A-T system is affected by electromagnetic devices, such as powerful stereo speakers being placed too close to the A-T electronic module in the vehicle's rear. I'm wondering if the same would apply to powerful external radio signals. There's a county emergency communications facility maybe 200 yards away.

The dealer's service department hasn't been very enlightening about what may be going on. In fact, the first couple of times I complained I was told it was normal for the A-T system to act as described above, if it had engaged for some reason. However, I think mine was probably the first one ordered by the dealer for stock, and I've only seen one other Explorer with the system show up on his lot.

This complaint is highly troublesome with respect to Explorer operation. This means:

- Ford has failed to correct basic GEM problems that were revealed by the 1999/2000 recall,[7] instead adding cheap and ineffective protective measures;
- The Explorer's wiring has been constructed so as to function as an efficient antenna for radio interference that is found in virtually all areas where the vehicle is used, and becoming increasingly prevalent in all areas;
- This susceptibility to radio frequency interference and Ford's failure to protect against it mean that new Explorers are perhaps even more prone to GEM failure than those built previously.

These user complaints, and others, document that the descriptions of Explorer technical problems described in this book are being experienced by owners and that Ford dealerships are not answering questions concerning these failures satisfactorily.

The protection of electronic components and systems against excessive power or surges is struggling to keep pace with evolving trends in semiconductor development. On one hand, some protection devices have become obsolete. Zener diodes have provided much of the existing protection, but their physical characteristics make them unsuitable for protecting circuits operating much under five volts. Power supplies may be the weakest component of a system. On another hand, decreasing feature sizes and interconnects have created new vulnerabilities to the ICs which make up the system.

These problems—even when considered separately from the issue of intentional disruption by radio frequency energy—require that better engineering techniques be used. Schottky diodes—invented in 1938—currently are used only in specialized applications where improved overvoltage protection is required. Their judicious use in devices such as computers and communications equipment will not add significantly to development or production costs, but will raise the threshold considerably for the power needed to disrupt these protected devices. Other techniques such as proper grounding and shielding will produce similar benefits at a small cost.

Our work with the Ford Explorer GEM system documented significant problems with the vehicle's operation. Among our findings:

- Analysis of the GEM module shows the addition of new features and functions with each new model vehicle. Each year shows changes and updates.
- To the present 2002 model, the GEM module has never been shielded for EMI or heat. Diodes have been put on the power sources to protect against over voltage or reverse voltage input along with clamp diodes for protection of semiconductor I/O ports.

---

[7] NHTSA Recall No. 00V072/Ford Recall No. 00S04

- Certain earlier models of the GEM module (1995 etc.) were affected by heat. The module was located under the dash near the radio. In cases where the hot sun heated the top dash area, some users complained that the electric windows would not work and the wipers would run after starting the vehicle. The vehicle worked normal after cooling and resetting the computer (turning off the ignition).
- Later models of the GEM modules were used to control the high/low shifting of the four-wheel drive transfer case (1995 and later). The GEM module would operate a relay that controlled an electric motor that would shift the transfer case from high speed to low, or vice versa.
- During testing of a 1998 Ford Explorer, it was found that by making certain wiring changes that simulated conditions a driver might experience, the transfer case could be shifted from high speed to low speed while moving. In testing, the transfer case was shifted multiple times at speeds including five and 35 mph (56 km/h).

6. Proposed Design Techniques to Counter Failure Mechanisms

The central point of this book is that incrementally designed and modified vehicle electronics systems are likely to fail. All elements of a vehicle's electronics must be designed with the harsh operating environment, and sometimes high consequence of failure, in mind.

That said, all integrated circuit (IC) chips are not created equal. Providing electronics for automobiles is one of the most challenging tasks for a semiconductor designer. In my experience, half the components within a chip designed for use in an auto will be specifically designed to protect the operational circuitry. It is also my experience that the automotive electronics market is perhaps the least forgiving from a cost-cutting perspective. Though much attention is focused on the problems raised in this paper, protection schemes that would be funded with no question in military acquisition programs, would have a hard time surviving in the extremely price sensitive world of the auto industry.

Designers often simply make mistakes, and the cost of redesigning even a simple chip runs into millions of dollars. Such a redesign effort is also time intensive, and there is no guarantee that the new product will be free of defects. A chip that has a logic flaw or a susceptibility to latch-up cannot be changed easily by an auto manufacturer, and there would be considerable pressure to use an expedient field repair rather than build a new unit. This perhaps was the case in the 1999-2000 GEM recall.

Circuit protection has long been a primary issue in the military and flight arena. If we examine the work being done, lessons can be learned that will directly apply to the automotive electronics industry. The intent of some of my earlier work was primarily to understand how electronic systems are affected by weapons that emit electromagnetic energy and how they can be protected against damage. Since microcircuits are the most vulnerable parts of modern electronic systems, much of my work has focused on the failure mechanisms of these devices and how to protect them against the high internal voltages that can be produced by radio frequency interference, static electricity, and other sources.

How the Strategy for Protecting CMOS Devices Is Different from that of Bipolar Transistors or P Channel or N Channel MOS Integrated Circuits

This analysis excludes the effects of CMOS latch-up effects, which will magnify the sensitivity to excess energy of any COTS integrated circuit or system. An analysis of the latch-up effects on the failure mechanism is the focus of a separate chapter. There is a difference in the requirements for effective protection from overvoltage occurrences between CMOS ICs and P Channel MOS (metal oxide semiconductor), N Channel MOS, or bipolar ICs. The key reason is the latch-up phenomena related to the CMOS structure. P Channel, N Channel, or bipolar ICs, if properly designed, do not exhibit latch-up characteristics and are therefore easier to protect from induced self destruction.

Modern design bipolar ICs usually operate from a five-volt (dc) power supply. As a result they can be protected against failure by a six-volt Zener protection diode, which is fast acting and rugged. The six-volt Zener can limit externally applied voltages even from low source impedances, to a relatively low voltage (five-to-six volts), thereby providing a good protection scheme.

While P Channel and N Channel high-power switching transistors are more robust in their resistance to overvoltage damage, the gate dielectric of the MOS devices can still be ruptured by unclamped high voltages. Similarly, the small geometry input transistors of bipolar devices can be damaged by high input currents if some form of voltage clamping protection is not provided. These voltage clamps can also be the same type of Schottky diode system (to be discussed later) as proposed for protecting the CMOS devices. In addition, a silicon Zener diode would also be adequate.

For the most part, the past strategies for any type of IC protection in COTS equipment was the use of a five-to-six volt Zener diode on all lines that address external system ports to prevent externally applied voltages from any source exceeding five volts. This is an effective method for most consumer products in the past, but it will not be effective in protecting the low voltage (3.3 and 1.5 volts) circuits that will be used in current and future products.

The key factor in developing effective protection from overvoltage occurrences between CMOS ICs and P Channel MOS, N Channel MOS, or bipolar ICs is protecting against the latch-up phenomena which is related to the CMOS structure. P Channel, N Channel, or bipolar ICs, if properly designed, do not exhibit latch-up characteristics and as a result are easier to protect from self destruction, or unexpected operations.

Latch-up and Strategies to Defend Against it

In the construction of complementary-symmetry metal oxide semiconductor (CMOS) Integrated Circuits (ICs), one must produce P Channel transistors (which require an N-type substrate material or N well) and N Channel transistors (which require a P-type substrate or P well). This is usually accomplished by starting with a P-type wafer (substrate) and forming deep N-type areas (wells) on the IC area in which the P-type transistors will be made. After this, the N-type source and drain elements (which form the N Channel transistor), and the P-type source and drain elements, are formed in their respective areas.

The end product, which contains millions of N and P transistors, allows very complex logic decisions to be accomplished. This is done by switching the outputs of the logic elements from the positive supply voltage to zero volts (ground). This occurs at a very high speed while consuming a very small amount of average current or power.

One of the limitations of this very efficient technology is that there is such a sufficient plurality of P-N junctions that an unstable state can be developed within various parts of the IC. This instability can result in undesirable high current paths between the positive supply voltage and ground. These current paths are the consequence of what is called "four-layer diode action" discovered by William Schockly in 1956.

Since the earliest days of semiconductor development, it has been known that if a structure is formed that consists of two P-N junctions, in series, on a single substrate, that this device can have two electrical states: a high impedance state or a low impedance state. The equivalent electrical schematic of the construction diodes of a CMOS IC is given in *Figure 39* and a cross-section diagram is given in *Figure 38*. In this construction, emitter one (E1) is formed by the P Channel transistor source, which is connected to the positive supply.

E2 is the drain of the P Channel transistor, which is often connected to an output pin—it is also the input protection circuit protection diode. This equivalent circuit is normally in the non-conductive state, causing little or no current flow from Q1 to Q2. When the voltage on E2 is increased to 0.7 volts above the voltage of E1 (+V power supply) a current flows from E1 through the base of Q1 and R1 to the positive power supply (+V). This current is multiplied by the beta (current gain) of Q1 and flows out of Q1's collector into the base of Q2, causing Q2 to become conductive. The base current of Q2 is multiplied by the beta of Q2 and the resulting Q2 collector current is extracted from the base of Q1, causing it to turn on harder and supply additional current. Since this circuit has a high positive feedback gain, it drives itself to a stable state of full conduction (saturation), where it remains until the power supply voltage is removed.

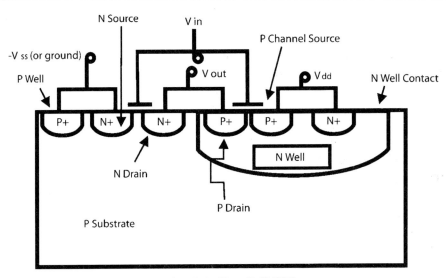

Fig. 38. **Cross-section of inverter circuit in N well CMOS**

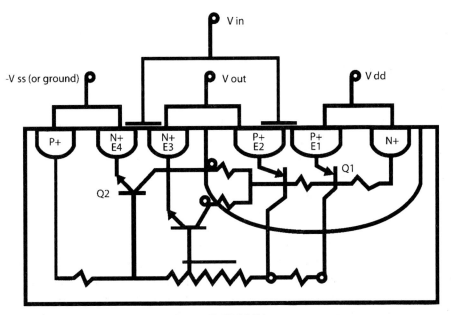

Fig. 39. **Parasitic bipolar portion of N well CMOS inverter**

A similar action takes place if diode E4 is taken negative to -0.7 volts and sufficient current flows from the circuit to establish the on condition.

The action relates to the CMOS IC in the following way. The complete CMOS IC is effectively an assembly of four-layer diodes resulting from its construction. In normal operation, the diodes are in the off or non-conductive condition. Within this structure millions of N and P MOS transistors operate reliably. However, if one of the P-type output transistor's drain is stressed to a voltage 0.7 volts above the supply voltage, this will force current flow into the IC four layer diode. This will trigger a portion of the IC to switch on (conductive state), causing a high current to flow through the IC.

Latch-up causes a temporary inoperative condition, at best, and often causes permanent damage. In an actual integrated circuit, Q1 and Q2 are made up of thousands of transistors in parallel. As a result, current can flow sufficiently enough to fuse (melt open) the one-thousandth of an inch diameter gold wires connecting the IC to the package pins. This current flow can also overheat portions of the IC to the point of destruction. If insufficient current flows, due to internal circuit resistance or resistance external to the IC, the latch-up condition will remain until the power supply current is removed for a short period of time (several micro-seconds).

During latch-up, under certain conditions, part of the IC will be non-functional and other portions functional, with some data lost when recovery is attained. When recovery is established, some portions of the chip may remain non-functional, but will not interfere with the usual tasks that are performed by the system. This partial damage can make evaluation difficult. There is also the possibility that some failures will heal within various time periods—hours or days or as a result of temperature cycling. The reverse is also true. Damaged, but functional areas may degrade and fail within various time periods—again, within hours or days or as a result of temperature cycling.

The interaction between printed circuit board (PCB) traces and equipment cables with high energy radio frequency will result in the appearance of voltages higher than the IC power supply at the IC signal input or output pins. Signals more negative than the ground voltage will have a similar effect to those that exceed the power supply positive voltage. If sufficient current flow occurs in the output pin, a latch-up condition will occur and there will be a circuit failure either temporarily or permanently.

The amount of damage caused as a consequence of this latch-up condition is dependent upon the supply current available from the power supply and the total circuit resistance. The total circuit resistance is determined by the chip architecture and what portion of the chip that is made conductive by the particular trigger current. It is, however, relatively simple to characterize any CMOS IC for latch-up sensitivity by exercising each pin on the IC with a DC voltage and recording the response.

The "on" state of an IC latch-up can be created by injecting or extracting a current into one of the P-N junctions on an input or output pin from an external source. The "off" state can be attained by temporarily removing the power supply current, as was the case in the 1999-2000 GEM recall. The added resistor reduced the chip latch current and allowed the chip to return to normal operation.

It would not be unreasonable to expect multiple pins to be affected simultaneously during exposure to excess voltage.

The indication of GEM lock-up was that the GEM computer would issue improper codes, which caused unrequested system operation. This was experienced by many owners when the windshield wipers turned on and could not be turned off, interior lights turned on and could not be turned off, and the four-wheel drive malfunctioned in various ways including a shift into four-wheel drive low unexpectedly. The nature of a lock-up on a CMOS chip is that it causes the chip to stop functioning, but allows the chip to assume a static condition in which the output signals (codes) can assume a random static condition until power is removed. The chip can then resume normal operation.

Methods have been developed to minimize the susceptibility of an IC to latch-up. These include: high conductivity substrates with epitaxial layers; output stage isolation; and Schottky diode protection of the output stage transistors. These measures can be quite effective, but they are often not used due to the additional costs of manufacturing the IC with this protection. The added cost results from the additional processing steps required. As a result, many commercial ICs and the equipment using them will be susceptible to electronic systems transients to varying degrees.

Currently Used IC Protection Techniques

There are a number of IC protection techniques currently in use. The primary objective is to protect against damage from electro-static discharge voltages. These are electro-static charges built up on a human body or an environmental surface due to movement of two or more static generating substances. This voltage is assumed to be 2,000 to 4,000 volts supplied from a capacitance of 20 to 200 Pico farads. This would be a typical charge on a human being. The discharge results in a low total energy shock provided from a high impedance source. This energy is sufficient to rupture the MOS gate dielectric, but does not usually cause CMOS IC latch-up due to its low total energy. This occurs primarily during assembly and test. Most input protection methods of ICs consist of some format current limiting resistor and two diodes, one connected to the positive supply and one to the ground. See *Figure 40*.

It is expected that the overvoltage will be conducted to the power supply or ground through the diode as it increases above the power supply voltage. The resistor limits the current that flows, thereby protecting the diodes from an overcurrent condition. The output pins of the IC are assumed to protect themselves since the drain diodes of the output driver transistors provide a similar function.

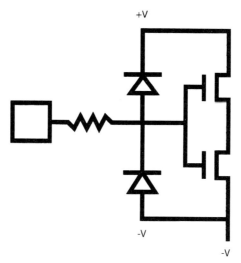

Fig. 40. **CMOS Input Circuit Showing Protection Diodes**

They have a much larger area and can dissipate more energy, eliminating the need for the current limiting resistor.

In recent years, experience has shown that these incorporated protection devices are not rugged enough to protect from higher energy discharges frequently experienced through use by general consumers. For this reason, COTS equipment makers have incorporated six-volt Zener diodes external to the ICs, to pins connected to external equipment ports. These diodes prevent higher voltage, and lower source resistance electrostatic voltages from exceeding +6 volts and -0.7 volts.

Because most of the ICs that have been produced in the past have used supply voltages of +5 volts, the 6-volt Zener protection device has provided adequate protection. The +5 volts is a low voltage limit for Zener diodes, and lower breakdown voltages are difficult to attain. Punch-through diodes can be made to work at lower voltage, but are costly to produce. As IC operating voltages approach 1.5 volts, a new protection device becomes necessary.

A New IC Protection Strategy

Now it has become necessary to reduce the IC operating voltage to 1.5 volts or less to reduce power consumption and to allow smaller IC dimensions to be used. A lower voltage protection device is now needed.

In the case where the detrimental influence is a high-energy electromagnetic impulse or high-frequency energy derived from the electromagnetic pulse, the characteristics of the damaging energy are completely different from low energy electrostatic voltages. This high-level induced energy can present high voltages with low source impedances at sufficient energies that are quite capable of damaging input protection diodes, output transistors, or producing latch-up conditions if the duration of the energy can be sustained over a sufficient period of time.

My proposed protection strategy for modern COTS IC products is as follows: It has been known for over 20 years that putting a low-forward-voltage Schottky diode in parallel with the IC diodes on the input or output pins of an IC will prevent latch-up caused by current flowing into these pins. These diodes are more robust and can protect IC inputs or outputs by limiting the voltage at these terminals. For unknown reasons, this protection method was never adopted by IC fabricators. It is still, however, a potent protection device and should be used to protect IC and equipment pins from high-energy fields. Because a metal/silicon diode (Schottky diode) has a lower forward voltage than a silicon P-N junction diode by a value of 0.1 to 0.2 volts, it will conduct current at a lower voltage. When the Schottky diode is placed in parallel with the IC diode it regulates the current flowing in the IC diode thereby preventing latch-up.

The switching speed of the Schottky diode is very fast and the leakage current is low—these are the other requirements of a protection device. The Schottky protection device connects diodes from the pin to be protected to the positive supply and ground. This arrangement clamps the voltage on the I/O (input/output) pin to within 0.5 volts of the power supply voltage. These devices, however, are not widely used in the market today, but could be developed and produced at a relatively low cost.

7. Conclusions

We spent some considerable time and effort taking apart vehicles to show that electronics designs in widespread use have unseen flaws that present users with hidden dangers. These flaws were not the inevitable result of pressing ahead with innovation, but rather the result of designers making some bad decisions. Ford Explorers were an obvious choice for study because of the emphasis on electronic innovation, but reviews of other vehicles would likely also turn up design problems.

The auto industry was responsible for more than 59,000 jobs in the electronics industry in the U.S. in 1998 (the most recent year for which such data is available), according to the Alliance of Automobile Manufacturers, and within the auto industry electronics is playing an ever-increasing role. According to a March 2004 Alliance statement:

> Cars today may have as many as 50 microprocessors, many of which make them easier to service. Every engine, every vehicle and every computer system is different—but all the sensors, all the output devices, must be in perfect "sync" for cars, minivans, trucks and SUVs to run efficiently ... The auto industry uses world-class technology in its vehicles. About 8% of the component value of the average U.S.-produced automobile is comprised of electronic content, representing a total of $12.4 billion in 2000 or about $970 per vehicle.

The benefits of vehicle electronics are tremendous, both in terms of high profit margins and market share for automakers. But the penalties for poor design work—high warranty costs, lost sales, and legal judgments—can be even greater. Chips and circuits that work well in the relative benign environments of the home or office will fail when subjected to the operational conditions imposed by vehicle use.

The incremental cost of designing and producing electronics that work well in the vehicle environments is often negligible or zero. No special materials or processes are needed in production. What is needed is a competent design approach, one that incorporates some or all of the techniques described in this book.

There is of course no substitute for testing and objective—hopefully independent—evaluation of design work. As the example of the errant clock mentioned at the beginning of this book illustrates, even talented engineers can make obvious mistakes. Marketing, product, and design managers must agree that new designs cannot be rushed onto the market in the hopes that they will please everyone and that nothing will go wrong. A lifetime is a long time, and professional and corporate reputations—not to mention the safety of customers—should be paramount.

APPENDIX: Author's Background

John H. Hall's career in semiconductor manufacturing began in 1962 when he was hired to help found Union Carbide's semiconductor operation. Since then, he invented the semiconductor design and process technology for a series of groundbreaking, successful commercial products, including: the first electronic watch; the first LCD digital watch; the first CMOS liquid crystal display hand-held calculator; the first electronic camera shutter, voice synthesizers; autofocus cameras; a low-power programmable heart pacemaker; and the first computerized heart pacemaker.

Hall also provided services to the U.S. Government for important new military technologies, including: a combination linear/digital low-cost sonobuoy IC; the phased array radar module for the B-1B bomber; the first radiation-hardened computer for a classified program; and a high-speed data acquisition system for a long-range infrared missile detection system.

Each of these commercial and military programs that Hall was involved in, consisting of inventing new solutions for electronics problems that had eluded other developers. Many of these solutions included making fundamental advances in semiconductor technology. For example, Hall invented the low-power CMOS technology that now forms the basis for virtually all of the consumer electronics products being produced today. A company he founded and led, Micro Power Systems, Inc., produced devices based on this technology for 10 years before it was adopted by Intel for use in its microprocessors and other products. This 40-year technical resume indicates why his companies have been able to make groundbreaking advances in speech recognition where other firms—including large ones such as Intel and Microsoft—have failed. His unique perceptions and insights related to analog/digital signal processing—technology at the center of the speech recognition challenge—have led to the significant advancements in the state of the art.

Hall is one of the foremost designers of groundbreaking integrated circuits, including those combining analog/digital technologies. Hall has used his product-development expertise to found companies and lead them to sustained growth. This extensive experience demonstrates Hall's ability to create new companies to exploit his technological breakthroughs.

Hall was co-founder of Intersil with Fairchild Eight member Jean Hoerni in 1967, heading its research and development, with work that included a breakthrough in coating silicon oxide gates with silicon nitride and creating the first practical N Channel metal oxide semiconductor (MOS) processes. Intersil also developed the first N Channel memory chip, which was later adopted as an industry standard.

Hall founded Micro Power Systems in 1971 with work that included low-power CMOS integrated circuit designs that he used in the first computerized programmable heart pacemaker, the first electronic camera shutter, the first low-cost ICs highly resistant to nuclear radiation, stationary phased array radar systems, frequency synthesizers, handheld digital voltmeters, hand-held LCD calculators, molybdenum gate MOS process used for cellular phone construction, and the first one-chip analog-to-digital converters. While he was president of MPS from 1971 to 1986, the company grew at an average rate of 25 percent a year, with no external debt or equity funding. From 1986 to the present, Hall has been president and CEO of Linear Integrated Systems, Inc., specializing in bipolar linear and high-speed CMOS digital circuits.

Below are some of the highlights of his development career:

1962: Named Union Carbide's Director of Integrated Circuit Development.

1962: Developed the first on-board aircraft computer made entirely of integrated circuits, used for the SR-71 Blackbird.

1963: Developed On-Chip Precision & High Value Thin Film Resistors.

1963: Conceived and Developed Dielectric Isolation ICs.

1963: Designed and oversaw construction of Union Carbide's first semiconductor plant, later to become Intel's first facility.

1964: Developed Monolithic Dual Transistors Using Dielectric Isolation.

1964: Developed FET Switch with SICR Thin Film Resistors for IBM.

1964: Developed Dielectric Isolation Video Amp with SICR Thin Film Resistors.

1965: Developed Dielectric Isolation μA709 with Super Beta Darlington Input for Litton.

1967: Developed Stepper Motor Quartz Watch IC for Omega and Gruen.

1967: Developed I2L Logic ICs.

1967: Developed N Channel Static RAM.

1967: Developed Automobile Clock IC for VDO.

1967: Co-Founder of Intersil with "Fairchild 8" Inventor Jean Hoerni.

1968: Developed Reliable and Commercially Reproducible N Channel IC Process.

1968: Developed Large Scale Integration Printing Calculator for Seiko.

1969: Developed Low Voltage CMOS LSI IC.

1969: Developed Bipolar 12 bit Digital-to-Analog Conversion Building Blocks for Analog Devices and Burr Brown.

1969: Developed the CMOS Quartz Wristwatch IC for Seiko .

1969: Developed Low Voltage Linear ICs for Heart Pacemakers for Cordis.

1969: Developed the I2L Pacemaker Logic IC for Cordis

1969: Developed as an in-house project the first electronic camera shutter chip, which was adopted by Canon for the first automatic exposure camera.

1969 Developed Custom High-Speed NMOS Static RAM for G.E. and Ampex.

1969 Developed High Speed N Channel ROM for Nixdorf Computer.

1970: Was principal architect of Seiko's first CMOS plant.

1971: Founder and President of Micro Power Systems.

1972: Developed CMOS Monolithic 10 bit D/A Converter with On-Chip Ladder Network for Analog Devices.

1972: Developed Super Beta Bipolar Linear Process for High Impedance Input Circuits.

1972: Developed CMOS with Two Layer Interconnect.

1972: Developed Commercially Reproducible Refractory Metal Gate CMOS.

1973: Developed Portable LCD Digital Voltmeter IC Chip Set for Dana, then Beckman.

1973: Developed CMOS Multiplexed Scanned LCD Calculator for Sharp.

1973: Developed Single Chip Integrating A/D Converter IC.

1974: Developed Low Power RF and Linear Pocket Pager ICs for Kokusai Electric.

1974: Developed Pocket Pager Chip Set, Digital Decode, and RF for Multi-tone, then for Harris Communications.

1974: Developed the CMOS Operational Amplifier Process (Including SICR Thin Film Resistors).

1974: Developed One Chip 14 Quad Slope A/D Converter for Analog Devices.

1975: Developed the world's first computerized programmable heart pacemaker for Medtronics after competing for the project with Motorola and Texas Instruments. Similar project completed for Intermedics.

1976: Developed One Chip CMOS 30 MHz Frequency Synthesizer with On-Chip Prescale.

1978: Developed Full Featured Linear BiCMOS Process (Including SICR Thin Film Resistors).

1979: Developed Bipolar Linear LSI Circuit Containing 14 OP Amps for Reliance Electric.

1980: Developed Monolithic 200 Volt Subscriber Line Interface IC for Philips.

1980: Developed On-Chip Capacitors with High Value 1 PF/MIL 2 Capacitance.

1983: Developed One Chip Linear BiCMOS Insulin Shock Monitor for Teledyne.

1984: Developed Bulk CMOS Flash A/D Converter, 6 bits at 30 MHz.

1984: Developed Bulk CMOS Flash A/D Converter, 8 bits at 20 MHz.

1984: Developed Low Cost Chip Assembly for B-1b Bomber Phased Array Radar Control Module Assembly for Westinghouse Defense Systems.

1984: Developed High Speed CMOS, 35 MHz Flip Flop Toggle Rates.

1985: Developed Highest Performance Amplifier and A/D IC Subsystem for Strategic Defense Initiative Infrared Focal Plane Sensor Array for Aerojet General.

1986: Developed CMOS One Chip Bell 212A Modem IC.

1986: Founder and President of Linear Integrated Systems, Inc. specializing in precision bipolar linear and high-speed CMOS digital circuits.

1988: Developed Four Layer Interconnect High Speed CMOS.

1989: Developed Zero Temperature Coefficient Thin Film Resistor Process.

1990: Patented CBiCMOS with Four Layer or Interconnect.

1991: Patented the Majority Carrier Injected CMOS.

1997: Developed and produced a field effect transistor with linear amplification properties more similar to those of a vacuum tube than any other semiconductor.

1999: Developed a next-generation hearing aid amplification IC.

Bibliography

Berliant, Adam. *The Used Car Reliability and Safety Guide*, 2nd ed. Cincinnati, OH: Betterway Books, 1997.

Bonnick, Allan W. M. *Automotive Computer Controlled Systems*. Woburn, MA: Butterworth-Heinemann, 2001.

Consumer Guide 4 x 4s, Pickups & Vans – Buying 2002 Guide. Lincolnwood, IL: Publications International, Ltd., 2002.

Consumer Reports – New Car Buying Guide 2000. Yonkers, NY: Consumer Union of United States, Inc., 2000.

Consumer Reports – Used Car Buying Guide 2002. Yonkers, NY: Consumer Union of United States, Inc., 2002.

Dammann, George H. *Illustrated History of Ford, 1903-1970*, revised ed. Sarasota, FL: Crestline Publishing, 1971.

Derrick, Martin. *50 Years of the Motor City*. London: PRC Publishing Ltd, 2002.

Ford Technical Service Bulletin, Bulletin No. 94 – 16. Plymouth, MI: Ford Motor Company, 1994.

Ford Technical Service Bulletin, Bulletin No. 96 – 24. Plymouth, MI: Ford Motor Company, 1996.

Ford Technical Service Bulletin, Bulletin No. 97 – 8. Highland Park, MI: Ford Motor Company, 1997.

Ford Technical Service Bulletin, Bulletin No. 98 – 26. Highland Park, MI: Ford Motor Company, 1999.

Ford Technical Service Bulletin, Bulletin No. 01 – 10. Highland Park, MI: Ford Motor Company, 2001.

Ford Technical Service Bulletin, Bulletin No. 01 – 13. Highland Park, MI: Ford Motor Company, 2001.

Ford Technical Service Bulletin, Bulletin No. 01 – 16. Highland Park, MI: Ford Motor Company, 2001.

Freeman, Kerry A. and S.A.E., eds. *Ford- Bronco II / Explorer / Ranger 1983-94 Repair Manual*. Radnor, PA: Chilton Book Corporation, 1994.

Jurgen, Ronald K., ed. *Automotive Electronics Handbook*, 2nd ed. New York, NY: McGraw–Hill, Inc., 1999.

1976 Truck Shop Manual, Volume I – Chassis. Dearborn, MI: Ford Marketing Corporation, 1975.

1976 Truck Shop Manual, Volume III and IV – Electrical and Body. Dearborn, Michigan: Ford Motor Company, 1975.

1991 Explorer Electrical & Vacuum Trouble – Shooting Manual. USA: Ford Motor Company, 1990.

1994 Ford Explorer Electrical and Vacuum Troubleshooting Manual. USA: Ford Motor Company, 1993.

1994 Truck Wiring Diagrams – Explorer FPS – 12240-94. USA: Ford Motor Company, 1993.

1994 S/T Truck Service Manual. Pontiac, MI: General Motors Corporation, 1993.

1997 Explorer Mountaineer – Electrical and Vacuum Troubleshooting Manual. USA: Ford Motor Company, 1996.

1999 Explorer Mountaineer Wiring Diagrams. USA: Ford Motor Company, 1998.

O'Donnell, Jayne. "Tech Advances Make for High-Priced, High-Class Headaches." *USA Today,* 11 November 2003.

Slavin, K., J. Slavin, and G. N. Mackie. *Land Rover - The Unbeatable 4 x 4.* Newbury Park, CA: Haynes Publications Inc., 1984.

Stidham, Todd. *Ford Ranger/Explorer/Mountaineer 1991-99 Repair Manual.* Newbury Park, CA: Haynes North America, Inc., 1999.

Storer, Jay, and J. H. White. *Ford Explorer Mazda Navajo & Mercury Mountaineer Automotive Repair Manual.* Newbury Park, CA: Haynes North America, Inc, 2001.

Storer, Jay and Haynes J. H. *Ford Explorer Mazda Navajo & Mercury Mountaineer Automotive Repair Manual.* Newbury Park, CA: Haynes North America, Inc., 2000.

2000 Explorer Mountaineer Wiring Diagrams. USA: Ford Motor Company, 1999.

Index